W9-CHB-917

OIL & GAS IN COMECON COUNTRIES

OIL & GAS IN COMECON COUNTRIES

Daniel Park

Kogan Page, London/Nichols Publishing Company, New York

First published in Great Britain in 1979 by
Kogan Page Limited, 120 Pentonville Road, London N1 9JN
ISBN 0 85038 135 5

First published in the United States of America 1979
by Nichols Publishing Company, Post Office Box 96, New York
NY 10024

Library of Congress Cataloging Card No 78-65910
ISBN 0-89397-040-9

Printed in Great Britain by Anchor Press, Tiptree, Essex
and bound by William Brendon and Son Limited

Contents

Acknowledgements

During the time spent in preparing this study I have benefited from discussions with a number of people whose help, criticism and interest have been a continual encouragement.

Above all I owe a great debt of gratitude to Alec Nove, Professor of International Economics and Director of Soviet and East European Studies at the University of Glasgow. Professor Nove guided this study throughout with good humour and fruitful suggestion: it was a privilege to work with him.

I am grateful to Mr D J I Matko, Mr R A Clarke, Dr Hannes Adomeit, Dr V V Kusin and Dr L Sirc of the University of Glasgow, who have given generously of their time and knowledge in discussing sections of the study. I extend my thanks also to Dr Philip Hanson of the University of Birmingham, and Professor Peter Odell of the Erasmus University, Rotterdam, who both made helpful suggestions on a wide range of issues.

Special thanks are due to Dr R J Hay of Imperial Chemical Industries Ltd, who over a number of years' friendship has discussed varied problems arising in international oil and chemical markets. Not least, I benefited considerably from the help of Ethel de Keyser of Kogan Page Ltd, who made many useful recommendations as the study was being prepared for publication.

Responsibility for errors, omissions or tenuous interpretation is mine entirely.

Daniel Park
Glasgow, September 1978

Introduction

Alec Nove *Professor of International Economics, Glasgow University*

Dr Park's theme is a very important and topical one. The USSR has become the world's largest producer of oil and will soon achieve the same status for gas. In Siberia a vast storehouse of mineral riches is to be found, which, by the end of the century, could give a great accession of economic strength to the USSR, in a world in which energy may be in increasingly short supply. On the other hand, Siberian riches lie in remote, inhospitable and empty areas, which are very costly to develop, and so the immediate future is one of stringency, with the Soviet planners striving to meet the demands of the domestic users and their Comecon allies, while simultaneously trying to expand sales to the West for much-needed hard currency. The CIA has forecast an actual decline in Soviet oil production, with very serious consequences both for the Soviet balance of payments and for the trade plans of her Comecon allies; they would have to buy a great deal of oil from the Middle East, and would have to earn the hard currency with which to pay for it. It was being asserted in some quarters that the Soviet bloc would become a major oil importer in the eighties, and this was used to explain Soviet policy in the Middle East.

Dr Park was from the first sceptical about the CIA estimates, which are by now somewhat discredited. He also noted the impressive achievements and prospects of the natural gas industry. Nonetheless, it is clear that the next years will be a period in which supplies of fuel in the Soviet Union will be tight. They would be very tight today, were it not for the lag in the investment program for certain fuel-using industries. The Soviet press prints many articles on the need for stringent economies, and publicity is given to efforts to develop nuclear and also solar energy. It is only right to add that the Soviet planning system seems to be coping with the energy problem with considerable skill: measures for conservation, exploration, large-scale investments in new areas, have been reasonably effective, and contrast with the confusions attendant upon President Carter's attempt to have an energy policy. This is due partly to the nature of the Soviet political system (no car lobbies and the like), but also because the centralized planning system is at its best when there are big, strategic decisions to be taken in such priority areas as fuel. Long-term plans take anticipated shortages into account, resources can be made available for such enormous projects as the development of the oil and gas resources of North-West Siberia, the building of thousands of miles of railway line (the Baikal-Amur trunk line) in the wilderness of East Siberia, the creation of a network of pipelines, some of them running from Asia right across European Russia into Germany, Austria and beyond. It is possible that a start might soon be made on an even larger pipeline leading from West Siberia to the Pacific, with the help of Japanese capital. There are many serious weaknesses in the Soviet economic system, but they show themselves least in the area of planning fuel and power – though here too problems have arisen in coordinating the activities of numerous ministries and departments, and in supplying high-quality specialized machinery, some of which is being imported from the United States and elsewhere.

True, in earlier periods the Soviet planners were very slow in correcting their overdependence on solid fuels, and it must seem remarkable that they managed to avoid using their vast resources of natural gas for so long. But even communist planners can learn lessons from experience and from past errors.

It is clear that Dr Park is dealing here with questions which have a major significance for the world's oil markets, for Western Europe's future dependence on oil and gas supplies from the 'East', for the economic relations of Eastern Europe with the Soviet Union and the extent of the future integration of Comecon, for the Soviet balance of payments and its capacity to import from the West and for the rate of economic growth of the USSR. All this has great political as well as economic significance. Dr Park has used his thorough knowledge of the oil industry to good advantage, and the result is a well-documented and researched book, which, as the reader will see, is a most valuable guide to the Soviet Union's most strategically important industries.

Alec Nove

Tables

Chapter 4

Chapter 5

Chapter 6

Chapter 7

Chapter 8

Notes

Measures and Conversion Factors

The metric system has been adopted throughout this book. 'Tonnes' is used in preference to 'tons' or 'metric tons'. One billion equals 1,000 million.

Soviet energy statistics are often recorded in 'tonnes of standard fuel equivalent' (tsf), 1 tonne containing 7,000 calories per kilogramme. The conversion factors for Soviet data are as follows:

Crude Oil	1 tonne	= 1.430 tsf
Natural Gas	1 billion cubic metres	= 1.190 tsf
Hard Coal	1 tonne	= 0.820* tsf
Brown Coal	1 tonne	= 0.420 tsf
Oil Shale	1 tonne	= 0.325 tsf
Fuelwood	1 tonne	= 0.249 tsf
Peat (to 1960)	1 tonne	= 0.400 tsf
(from 1960)		= 0.325 tsf

**after 1970 this figure is 0.780*

The following corresponding factors are used for Eastern European countries:

Country	*Natural Gas*	*Crude Oil*	*Hard Coal*	*Brown Coal*
Bulgaria	1.20	1.40	0.59	0.29
Hungary	1.20	1.40	0.71	0.50
GDR	0.47	1.40	0.71	0.30
Poland	1.21	1.40	0.80	0.29
Romania	1.36	1.40	0.71	0.36
Czechoslovakia	1.26	1.40	0.80	0.48

(other factors, as for USSR)

Sources for the derivation of these factors are given in J Bethkenhagen, *Bedeutung und Möglichkeiten des Ost-West-Handels mit Energierohstoffen*, (Deutsches Institut für Wirtschaftsforschung, Sonderheft 104) Berlin, Duncker & Humblot, 1975, p 102 and R E Ebel, *Communist Trade in Oil and Gas*, New York, Praeger, 1970, p xix

Transliteration

The cyrillic transliteration scheme adopted by the British Museum and British Lending Library has been used, but the following exceptions have been made: final ий is 'ii', final ый is 'yi'; щ in Bulgarian is 'shch'; ъ in Bulgarian is 'o', as the hard sign in Russian is represented by ". Where an author's name appears in a language other than his own it is represented in that language or transliterated in accordance with the scheme. Instances of this are few.

Chapter 1

The Development of the World Petroleum Market, 1960-1975

In the years between 1960 and 1975, but particularly after 1970, the world petroleum market changed dramatically. North American, Japanese and West European industry were enjoying a high rate of growth at this time, due largely to the availability of low-cost oil for import from Middle Eastern and other countries, and later due to the indigenous production of natural gas in the US and Western Europe. The apparent ease with which a small number of highly integrated multinational companies controlled exploration and production, and to an extent the processing and distribution, of oil, contributed to the widespread mood of confidence among government and industrial circles in these growing economies that the availability of cheap hydrocarbon energy would continue. However, during this time the countries in which the multinational companies operated brought to fruition a policy of their own, intended to curtail activities which they believed to be against their long-term interests. These activities included the rapid development of oil (and in some cases gas) reserves at such a rate that oil prices and revenue per barrel to the producer countries declined.

Throughout the sixties the oil companies engaged in keen competition for an increasing share of the expanding markets, primarily that of Japan and Western Europe. The producer countries, on the other hand, had been accumulating economic and political strength ever since the formation of the Organization of Petroleum Exporting Countries (OPEC) in 1960, and in the latter part of 1973 and the first few months of 1974 they independently raised the selling prices of their oil and decided on production policy – thereby achieving two of OPEC's prime objectives. The process behind these decisions and the extent of their impact secured a fundamental shift in the power structure of the world petroleum market: the driving force passed from the oil companies to the producer governments acting through OPEC. Some five years have passed since the Yom Kippur War and the 'oil crisis', and the power of OPEC shows no sign of waning.

During this period the Soviet Union appeared to remain insulated from the events that shaped the development of the world market. Once the immediate problems of post-war reconstruction were over, the Soviet Union, exhibiting an impressive rate of economic growth, first regained self-sufficiency in energy and then became once more an exporter of oil, as it had been in the thirties. The rapid rise in energy demand, especially for oil, in the relatively conveniently located markets of Western Europe coincided with the Soviet Union's development of substantial reserves of coal and natural gas (the latter at a comparatively late stage by world standards) as well as the exploitation of the prolific Volga-Ural oilfields. Exports of oil and refined products showed steady growth from the early sixties. It seemed likely, therefore, that the Soviet Union would be well placed to continue to satisfy its own demand for energy, to accelerate the trend from coal to hydrocarbon fuels within the domestic energy balance, to meet the growing demand in the member-countries of Comecon* – who, with the exception of Romania, were and remain poorly endowed with hydrocarbon reserves – and to continue to take advantage of opportunities for hydrocarbon exports to the hard-currency markets.

However, by the late sixties there arose some measure of doubt in the West as to the Soviet Union's ability to maintain the rate of development that had been achieved in the late fifties and early sixties. There was evidence that the development of the Volga-Ural oilfields was running into serious problems, and most enterprises operating in that region were facing escalating costs. The rate of growth in oil production and in exports was declining and the Soviet Union began to negotiate with certain OPEC countries for the supply of crude oil, and advised the Eastern European countries to do likewise. The view was expressed in a number of quarters that the Comecon bloc might become a competitive bidder for oil produced by the Arab-American oil companies, whose internal relations were becoming increasingly unstable.

The development of the world petroleum market can be viewed as being dependent on the interaction of four major spheres of influence. First, there are the large international companies, whose prime objective is long-term profit maximization and who operate across national boundaries, optimizing profits within international legal and economic constraints. Second, the home governments of these companies, who have become energy importers, need to secure energy supply at the lowest cost to the national economy. They have also to bear in mind the international consequences, political and economic, of their policy decisions since they are historically the major participants in international financial and political institutions. Third, the Organization of Petroleum Exporting Countries (OPEC) seeks through its collective ownership of a single vital commodity to raise its member-countries out of economic backwardness. In the light of the deteriorating socio-political relations in the Middle East it seeks also to found a new order in which these countries can enjoy a greater measure of influence. Fourth, the Soviet Union, self-sufficient in energy but facing a number of problems as the economy develops that might be solved only by increasing trade and cooperation with the West, is committed to a policy of, at least, containing the spread of Western influence.

The Influence of Oil and Gas Companies

The international oil market is dominated by seven large integrated companies[1] which have often joined together in exploration and production activity in the Middle East, and in negotiating concessions with governments on whose territory their operations are sited. Although accounting for the greater part of the

* Throughout this study *'Comecon'* denotes the USSR, plus Poland, Romania, Hungary, the GDR, Czechoslovakia and Bulgaria.

non-communist world's oil production, the major companies have met with increasing competition, attracted in the first instance by the comparatively high profitability of the industry. From 1953 to 1972 over 300 private and 50 state organizations either entered the international oil industry or expanded their existing operations.[2]

The world oil market – which, given the self-sufficiency of the Comecon bloc and the effect of oil import quotas imposed in the US in 1959[3], meant essentially Western Europe and Japan – was highly competitive during the sixties. This, coupled with the re-emergence of the Soviet Union as an oil exporter and the discovery of natural gas in the Dutch Groningen field in 1959, led both to the decline of the European coal industry and to a high level of dependence in Western Europe on imported energy.[4] The attractiveness of the Western European market and its competitive nature gave rise to declining profitability in the operations of the major companies. This latter fact generated some doubt as to the continuing capacity of the Western oil industry to finance itself, and led to the formation in 1968 of the London Oil Policy Group.[5]

The Group consisted of representatives of the major oil companies and met fortnightly. It drew up a strategy based on the continued availability of oil from the Middle East at a price which would rise gradually, relative to the general commodity price-level in Western Europe, where the market for oil and refined products was planned to grow in an orderly and predictable manner. The price rises imposed by Libya's Colonel Gadafy in 1970, and the agreements of OPEC in Tehran and Tripoli in 1971, and in Geneva in 1972, were initially neither against the interests of the companies nor against their wishes. On the assumption of an expanding market the major companies continued their investment programs expanding their facilities and gearing the level and timing of investment to anticipated improvements in the financial state of their operations. At no stage did the companies envisage relinquishing control over decisions determining crude oil prices or production levels.

This strategy was seriously undermined when OPEC assumed control in 1973-1974, accomplishing in a few months what the oil companies had sought to accomplish over years. However, one direct result of higher prices for OPEC oil, imposed at a time when no ready substitute existed, was a reduction in the medium-term demand for refined products. OPEC's pricing policy worsened the economic depression of consumer countries and prompted consumer governments to impose restrictions on the use of oil in order to offset a potentially serious deterioration in their balance of payments.

The structure of the international oil industry underwent marked changes in the postwar period. Profit attracted competition and the industry grew from a highly concentrated process involving fewer than 10 participants to a network in which 50 or more integrated and, by world industrial standards, large companies cooperated and competed.[6] The seven major oil companies, whilst maintaining impressive rates of growth, lost market share to competitors attracted by their success. This decline was reflected differently in each sphere, where newcomers to the industry saw the best opportunities for a return on their particular resources (see table 1.1). A comparative analysis of the foreign capital expenditure of the seven major companies (see table 1.2) reveals the changing pattern of the

Table 1.1 Analysis of Structural Change in the International Petroleum Industry 1953 and 1972 (% share)

Sphere of Activity	*1953*		*1972*	
	7 Majors	*Others*	*7 Majors*	*Others*
Concession Areas	64	36	24	76
Proven Reserves	92	8	67	33
Production	87	13	71	29
Refining Capacity	73	27	49	51
Tanker Capacity	29	71	19	81
Product Marketing	72	28	54	46

Source: N H Jacoby, *Multinational Oil*, New York, Macmillan, 1974, p 211

Table 1.2 Total Foreign Capital Expenditure of Capitalist Oil Companies 1948-1972 ($ billion, % of the total)

Year	*7 Majors*	*%*	*Others*	*%*	*Total*
1948	1.2	75.0	0.4	25.0	1.6
1950	0.7	46.0	0.8	54.0	1.5
1955	1.1	37.9	1.8	62.1	2.9
1960	1.7	32.1	3.6	67.9	5.3
1961	1.7	32.1	3.6	67.9	5.3
1962	1.9	35.2	3.5	64.8	5.4
1963	2.0	35.1	3.7	64.9	5.7
1964	2.3	37.1	3.9	62.9	6.2
1965	2.8	41.2	4.0	58.8	6.8
1966	2.9	38.2	4.7	61.8	7.6
1967	3.1	39.2	4.8	60.8	7.9
1968	3.5	36.5	6.1	63.5	9.6
1969	3.8	37.3	6.4	62.7	10.2
1970	4.1	34.5	7.8	65.5	11.9
1971	5.1	35.2	9.4	64.8	14.5
1972	4.6	28.9	11.3	71.1	15.9

Source: *ibid*, p 250

industry's financial structure.

Given the US import restrictions, the companies' main commercial opportunities were to be found in Western Europe and Japan. One effect of the competition for the Western European market was a significant fall during the sixties in prices for refined products (see table 1.3). As a result of these relatively low energy costs, the Western European manufacturing industry enjoyed an advantage in world markets. Pressure was brought to bear on the US government by US energy-intensive industry to encourage the oil companies to negotiate oil price rises.[7] It has been argued that this constituted an 'unholy alliance' between the US government, the major oil companies and the OPEC Secretariat, to undermine the success of Western European industry.[8] A view also held is that the 'alliance' struck was between OPEC and the oil companies, to transfer the process of profit-generation from the oil-producing states to the consuming countries.[9]

It does indeed appear that OPEC has been strengthened and the long-term profit potential of the oil industry enhanced at the expense of Western European industrial consumers. But this analysis of the role of the oil companies and suggestions of an 'alliance' can easily overlook the fact that while the rise in the price of OPEC oil during the seventies was planned, it was not necessarily intended to have so severe an effect: oil was used as a (highly effective) weapon in a long-standing conflict, but probably without appreciation of the long-term consequences. Certainly neither OPEC nor the major oil companies sustained any economic loss as a result of oil policy in the seventies, but the 'alliance theory' must be tempered by consideration of the various factors that diminished the companies' influence.

The significance of the Libyan price rise of 1970 was that it was imposed *unilaterally* by Gadafy and not via the negotiating process set up by the London Oil Policy Group: Gadafy's action 'forced a settlement on the companies that irrevocably broke the pattern of the past'.[10] There was no question of isolating Libya within OPEC in order to maintain the effectiveness of the Group; demand was high and the tanker market no longer as flexible as it had been in 1967 to cope economically with the loss of Libyan supply.[11]

After 1970 the Group's effectiveness in the market waned, although the trend in OPEC's pricing policy did not initially run counter to the Group's interests. The OPEC meeting held in Caracas in December 1970 cast further doubt on the 'alliance hypothesis': one particular resolution laid down minimum acceptable terms for increases in the wellhead prices of oil within a specified time scale. These terms also provided for enforced compliance by means of united action, should negotiations with the companies prove ineffective.[12] It has been argued that in the aftermath of the price rises of 1973 and 1974, companies became passive participants, having learned 'to accept defeat . . . and to roll with the punches'.[13]

The companies have maintained their financial well-being. Their influence, however, has been markedly diminished and their dependence on an insecure source of supply has prompted greater interest in exploration and development in areas other than the Middle East and South America. These areas include the Soviet Union.

Table 1.3 Average Annual Prices of Refined Product Imports in West European Markets 1957-1970 ($ per barrel)

Importing Country	*1957*	*1958*	*1959*	*1960*	*1961*	*1962*	*1964*	*1966*	*1968*	*1970*
Belgium	5.11	4.45	4.06	3.75	3.57	2.91	3.07	2.81	3.16	3.27
France	5.21	5.35	4.91	4.92	4.90	4.64	4.22	4.10	4.12	4.05
West Germany	5.11	4.22	4.33	4.13	3.99	3.67	3.37	2.82	3.49	3.24
Italy	6.22	7.05	4.25	3.45	3.33	2.87	3.48	2.93	2.93	3.65
Netherlands	5.66	4.05	3.69	3.53	3.61	3.49	3.29	3.02	3.72	3.54
UK	5.55	4.65	4.54	4.05	4.22	3.05	3.61	3.26	3.80	3.39

Source: Jacoby, *op cit*, p 239

The Influence of the Governments of Consumer Countries

The US government can be said to be the most directly involved in international oil in that it is the home government of five of the seven major companies. However, Britain, as the home base of BP and the center of the world coordination of Shell, despite predominantly Dutch influence, also plays a leading role. Government aims can be usefully viewed as those of securing availability of low-cost energy supplies, and of minimizing the negative impact of energy imports on the balance of payments and, particularly in the case of the US, on foreign policy options. There has been a marked growth in dependence on the part of each of the major economic powers with the exception of the Soviet Union (see table 1.4).

Table 1.4 Dependence of Major Economic Powers on Imported Energy (%)

	1955	*1960*	*1965*	*1970*	*1974*
EEC (9)	19	30	46	59	62
Japan	23	41	65	84	88.5
USSR	–	–	–	–	–
USA	1	6	7	8	19

Source: D F Cooper, *Energy: A Matter of Interdependence*, Conference Paper, Institute of Purchasing and Supply, 1976, p 3

Japan has been dependent on imported energy for well over the 20 years represented in the table. Domestic production has been severely limited by lack of resources and the 'economic miracle' has been substantially facilitated by low-cost imports. Table 1.5 shows the extent of Japanese dependence on external sources of oil in the period up to 1975. In the aftermath of the 1973-1974 price rises Japan was quick to oblige industry to export to the newly-rich OPEC bloc and to take steps towards developing trade links with non-OPEC oil producers, notably the Soviet Union and China.

At the same time a government report stressed the need for the Japanese oil industry, till then confined to refining and distribution, to involve itself to a

Table 1.5 Japanese Oil Production and Consumption 1968-1975 (million tonnes)

	1968	*1969*	*1970*	*1971*	*1972*	*1973*	*1974*	*1975*
Production	0.7	0.7	0.8	0.7	0.7	0.7	0.7	0.6
Consumption	142.7	169.0	199.1	219.7	234.4	269.1	258.9	244.0

Source: *BP Statistical Review of the World Oil Industry 1976*, pp 18, 20

greater degree in exploration and production. This was stressed on the grounds that widening the scope of the industry might strengthen the country's negotiating position in relation to OPEC.[14] The immediate effect of the oil crisis on Japan was that in a year of 11 per cent growth the fourth ('crisis') quarter showed only 1 per cent growth. The Japanese Economic Planning Agency estimated a maximum of 2.5 per cent growth for 1974, this following a 20-year period in which Japanese GNP grew at between 8 and 12 per cent per year, energy consumption by a total of 500 per cent and oil consumption by 2300 per cent.[15]

Despite the country's soaring demand for oil the Japanese government was able to devise a short-term energy policy which within two years of the 1973 crisis had minimized the impact of the price rises (see table 1.6). In 1973 Japan had reached the point at which oil accounted for over 70 per cent of energy requirement and 99 per cent of it was imported. Some 75 per cent of refined products were consumed by industry.[16] Given the extent to which the country's industry is dependent on oil, Japanese technical and financial involvement in oil developments currently extends to Canada, West Africa, the Middle East, Australasia, the North Sea, China and the Soviet Union.[17] In the case of deliveries from China, the first sale consisted of 1 million tonnes of crude oil from the Taching field in 1973, which led to the conclusion of a contract in February 1974 for the delivery of 1.5 million tonnes in that year.[18]

Table 1.6 Japanese Oil Consumption in Fiscal Years 1971-1975*

	1973	*1974*	*1975 1st half*	*1975 2nd half*	*1975*
Crude Oil Imports (million barrels)	1,814.6	1,760.0	805.1	866.9	1,672.0
% Change	+ 17.2	− 3.0	− 9.9	+ 0.1	− 5.0
Cost ($ million)	8,673	19,997	9,400	10,600**	20,000**
% Change	+ 108.8	+ 130.6	− 5.5	+ 5.4	no change

* *Fiscal 1973 ends 31 March 1974, etc*
** *Estimates, excluding October 1975 OPEC price rise*

Source: *Petroleum Economist*, January 1976, p 4

The origins of the US's energy problems are complex. Since 1970 oil production has declined whilst demand has grown rapidly (see table 1.7). With the exception of Alaska (which has not yet played a significant part in production), discoveries of new oil reserves have been unimpressive. The refining sector has been operating almost at capacity and production of natural gas, whilst increasing, has not kept pace with growth in demand. In 1967, when the Arab members of OPEC declared an embargo on deliveries to the US, there was scope for diverting Middle East oil to other destinations and re-exporting refined products. There was also the opportunity of substituting oil from certain countries not participating in the embargo, notably Iran, to counteract the loss. At the time underutilized refinery capacity provided the flexibility required to cope with differences in oil-processing characteristics, where these occurred.

Table 1.7 Production and Consumption of Oil and Natural Gas Liquids in the US 1960-1975 (million tonnes)

	1960	*1965*	*1970*	*1971*	*1972*	*1973*	*1974*	*1975*
Production:								
Crude Oil	350.3	387.6	478.6	469.9	470.1	457.3	436.8	415.9
NGL	33.8	43.6	58.9	60.1	62.1	61.7	59.9	58.0
Total (A)	384.1	431.2	537.5	530.0	532.2	519.0	496.7	473.9
Consumption (B)	473.0	549.0	694.6	719.3	775.8	818.0	782.6	765.9
C. A as % of B.	81.2	78.5	77.4	73.7	68.6	63.4	63.5	61.9

Source: *BP Statistical Review of the World Oil Industry 1970*, pp 18, 20; *1976*, p 18, 20

By 1970 this flexibility had been lost. The first problem to manifest itself was a shortage of fuel oil. In the first six months of 1970 demand for fuel in the US was 14.9 per cent above that for the corresponding period in 1969.[19] The industry was unable to compensate because artificially low 'ceiling prices' set by the government had discouraged investment in the industry. Gas consumption was close to production capacity, and demand was expected to remain high since the effect of rising costs in the domestic oil industry, coupled with the oil import quota system, was to force up prices of refined products. One analyst has put forward the view that ceiling prices were imposed in order to 'protect consumers against exploitation at the hands of the natural gas (pipeline) companies'.[20] It is more likely, however, that there has been a politically influential lobby of gas consumers enjoying a relatively low-cost fuel, who still favour the maintenance of such prices.[21] The US gas industry therefore recorded low profitability, an absence of investment incentive and a deteriorating ratio of reserves to production. Companies involved in the production and distribution of gas were prompted to seek additional supplies from outside the US, including the Soviet Union.[22] These attempts were not expected to make any impact on the short-term supply position in the US gas industry. Even if the Soviet-American joint projects currently under discussion do come to fruition, their contribution to the US energy supply will not be felt before the early eighties. Table 1.8 shows the development of the US gas industry from 1950 to 1975.

Accustomed to near self-sufficiency in energy, to continuous economic growth and to the role of economic and political superpower, the US government launched 'Project Independence 1980' in the immediate aftermath of the 1973 Middle East crisis. The objective was to regain self-sufficiency in energy. Some two years later it was estimated that by 1980, approximately 40 per cent of US oil demand would be met by imports from the Middle East and North Africa.[23] The Director of the US Federal Energy Administration eventually admitted that the economic loss that would be incurred in attempting to fulfil the requirements of 'Project Independence' was acknowledged to be more severe than the potential disruption of supply.[24] In the oil sector, as in gas, but for different reasons, diversification of the source of supply had become a pressing issue.

Table 1.8 The US Natural Gas Industry 1950-1975 (billion cubic metres)

Year	*Marketed Production*	*Imports*	*Consumption*	*Proved Reserves (Year End)*
1950	178	–	171	4,208
1955	266	0.3	257	6,300
1960	452	4.4	354	7,428
1965	454	13	454	8,112
1970	621	23	624	8,233*
1971	637	26	642	7,895*
1972	638	29	652	7,354*
1973	641	29	650	7,078*
1974	612	27	626	6,715*
1975	569	28	na	na

**Including Alaska (74 billion)*
na = not available

Source: *Petroleum Economist*, March 1976, p 85

Cheap energy was the base on which the Western European industry grew in the fifties and sixties. This was mainly in the form of imported oil, the price of which declined substantially during the sixties. As early as 1972 Odell stressed Europe's vulnerability in the face of increased taxes imposed on the oil companies, the companies' ability and intention to pass these on to consumers and the possible disruption of supply in the event of negotiating difficulties.[25] These difficulties would be considerable, he argued, if the price of OPEC oil was to rise to four dollars per barrel by 1975.[26] At this time also, the Chairman of the British National Coal Board, Derek Ezra, sensed the growing difficulties that might be caused by the lack of cohesion in US energy policy and drew attention to the possibility of the US and Western Europe competing for oil that could be subject to supply restrictions.[27] Ezra outlined Western Europe's dependence on energy imports which greatly exceeded that of the US and was likely to grow more rapidly, given the higher general rate of industrial growth compared with that of the US.[28] Bearing in mind also the increasing demands of Japan and the Third World, Ezra, like Odell, argued the case for a reduction of Europe's dependence on imported energy and advocated (understandably) investment in the coal industry, on the grounds that coal would be fully competitive with oil and gas in the eighties.[29]

In contrast, Odell argued the case for a Western European energy economy based on indigenous gas, there being in his view no opportunity in the short-term to exploit oil reserves at a level sufficient to have a marked effect on the energy balance.[30] He noted that as long as Middle East oil production costs remained at around 25 cents per barrel, there was scope for OPEC to revise prices downwards and to increase output.[31] He suggested, on the basis of an argument put forward by the American energy economist M A Adelman, that there was no physical or economic factor which might rule out such a possibility.[32] As to government reaction if faced with this problem, his position was that they should 'avoid . . . further entanglement with such insecure oil' and 'leave it for the poor, oil-importing countries of the Third World, where . . . cheap energy is much more important for ensuring development'.[33] Bearing in mind the pressures brought to bear on the US government by energy-intensive industry to ease import controls when faced with increasing competition in world markets, it is hardly likely that a West European government would impose other than a temporary tariff or quota system, since, as the US discovered, the overall economic loss from such a policy can be considerable.[34]

An OECD report, published in 1974,[35] reassessed the Organization's forward energy requirement in the light of the 1973-1974 oil supply crisis, underlining the economic difficulties that a policy aimed at self-sufficiency would cause. The direction of investment into developing energy reserves possessed by certain OECD members would, the report indicated,[36] limit the member-countries' prospects for economic growth. The argument put forward was that the major element of energy strategy should be to decrease the energy-intensity of the economies of member-countries as a whole. Such a policy would in theory alleviate some of the impact of the oil price rises but would not decrease the level of dependence.

The problem faced by the EEC and OECD in attempting to determine a common energy policy has been the intrinsic disparity of objectives and opportunities amongst their members. For example, Britain seeks to maximize the benefits of North Sea oil and gas in relation to her balance of payments problems and the government, at times in conflict with the industry, wishes to retain a measure of independence from EEC and OECD bodies in determining depletion and trade policy. France seeks to achieve as high a level of energy autarchy as possible, to the extent of non-participation in the 1974 Washington energy conference.[37] One feature which the EEC member-countries now have in common is substantial over-capacity in oil refining, expansion of which was undertaken prior to the 1973-1974 crisis. At that time it was anticipated that demand for refined products in Western Europe would continue to grow at the same rate as had been recorded in the sixties. Therefore one of the prime objectives of the oil companies operating in Western Europe is to employ these facilities more fully.[38] The possibility that the Soviet Union might prove to be an alternative to OPEC prompted several approaches by Western organizations.

The Influence of the Petroleum Exporting Countries

The third major influence in the world petroleum market is the Organization of Petroleum Exporting Countries (OPEC). From its inauguration in 1960 'as a defensive mechanism to form a common front vis-a-vis expatriate oil firms and major oil-importing countries',[39] with its objective 'not (the) regional integration of members' economies and societies, but mainly the coordination of members' policies in one commodity, petroleum, and solely in the export market',[40] OPEC sought throughout the sixties to maintain oil prices 'with due regard to the interests of securing a steady income to the producing countries, an efficient, economic and regular supply of this source of energy to consuming nations and a fair return on capital to those investing in the petroleum industry'.[41] It has been pointed out that relations between OPEC, the oil companies and their domestic governments are most appropriately regarded as a bargaining process rather than one of trade and commercial competition.[42] Study of the changing role of OPEC in the world market shows the shift in emphasis from a preoccupation with prices in 1960 to a desire for self-determined growth via complete ownership and control over output and pricing decisions.[43] The event that confirmed the turning point in the bargaining process was the successful imposition by Libya's Colonel Gadafy of production cutbacks, increased port charges at Libyan terminals and, effective from 1 September 1970, an increase of 30 cents on the price of $2.23 per barrel, plus increased royalty rates.[44]

In response to Gadafy's unilateral success an extraordinary meeting of OPEC, held in Tehran in February 1971, imposed price rises averaging 33 cents per barrel on crude oil delivered from the Persian Gulf.[45] The following month, at a further meeting, prices for crude oil delivered from Mediterranean terminals were raised proportionally to the Gulf prices.[46] Subsequent meetings of OPEC in January 1972 and June 1973 led to additional price rises to compensate for the devaluation of the US dollar.[47] Despite these increased prices, demand for Middle East oil continued to grow and investment decisions were made by the oil companies that provided for the expansion of refining and distribution in Western Europe.

Prior to October 1973, in negotiations between the producer countries and the operating companies, attempts were made to reconcile differing objectives, though Western inflation and the falling value of the dollar strained the agreements concluded. The Yom Kippur War quickly fused the economic and political dimensions of international oil trade, as OPEC abandoned the will to negotiate and exercised the oligopolistic power it had gradually acquired.[48]

In addition to the aforementioned impact on demand in consuming countries, the oil price rises confirmed OPEC as the driving force in the world petroleum market and brought about a radical shift in world distribution of wealth largely at the expense of industrialized nations, though this did, of course, work to the disadvantage of many of the oil-importing developing countries. This transfer of wealth conferred on OPEC members a substantial level of flexibility, a fact which has not escaped the notice of Soviet policy-makers.

There had been some optimism that the international system could again be brought into balance by recirculating oil revenues through Western financial institutions, thus easing currency transfer problems and providing through OPEC's increased purchasing power the means by which the health of Western economies could be preserved. It has been pointed out, however, that many of the OPEC member-countries do not need high levels of Western capital, since they are not experiencing a revolution of rising expectations.[49] The Soviet response to OPEC's enhanced financial power has been to sound repeated warnings that the petrodollar recycling process merely contributes to the capacity of certain Western countries to support the Israeli cause. A recent feature of discussions between the Middle Eastern member-countries of OPEC and the Soviet Union has been the negotiation of dollar loans out of surplus oil income and the possibility of Arab investment in the Soviet bloc.[50] This may become more attractive to the Soviet Union, given that questions concerning Soviet and East European indebtedness are currently being posed in some quarters of the Eurocurrency market.

As long as OPEC maintains its solidarity, it enjoys the status of a superpower in the economic sense. There are differing objectives within OPEC. The 'industrialized' members, such as Iran, Iraq and to a lesser extent Nigeria and Algeria, are interested in rapidly acquiring hard currencies that can be used to purchase goods and services to facilitate economic growth. Consequently their preferred strategy is to maximize the short-term inflow of such currencies. The 'desert state' members, however, in whose countries the bulk of oil and gas reserves are located, have small populations, few opportunities for industrialization, apart from petrochemicals, and a strong interest in the maintenance of political influence as long as the Arab-Israeli conflict remains unsolved. Their strategy is to secure long-term consumption of oil, by avoiding a pricing policy that would lead to an undesirably high level of development of alternative energy sources.[51] It is admitted by the Saudi Minister of Petroleum that OPEC has difficulty in determining and sustaining a unified policy.[52] However, member-countries realize that unity confers strength: the strength of OPEC is the breadth of its choice of strategy.

The Influence of the Soviet Union

The development of oil and gas in the Soviet Union and Eastern Europe is the fourth major influence on the world petroleum market. The trend from coal to hydrocarbon fuels took place at a later stage in Comecon than elsewhere in the industrialized world. The expansion of the oil industry commenced in the fifties, that of the gas industry in the sixties. During this time the Soviet Union resumed exports of oil and commenced exports of gas not only within Comecon but also in world markets.

The ninth Soviet Five-Year Plan for the oil and gas industries had to take account of the fact that reserves were predominantly located in Eastern areas. The same is true, though to a lesser extent, in the coal sector. According to a Soviet estimate published in 1975, some 90 per cent of the total energy reserves are located east of the Urals, whereas 70 per cent of consumption is in European Russia.[53]

It was intended at the outset of the ninth Plan that Siberia would account for the major part of the increase in Soviet oil production: in fact Siberian operations had to overfulfil their original objectives in order to compensate for production shortfalls elsewhere. In the gas industry production increases were scheduled for each area, including a substantial expansion of Siberian operations. As a result of a general shortfall against the original Plan, the Siberian share of Soviet gas production recorded in 1975 was as envisaged in 1970, when the Plan was being formulated.

With the exception of Romania, the Eastern European countries are poorly endowed with oil and gas reserves and have hitherto been dependent on the Soviet Union as almost their sole supplier. The trend to hydrocarbon fuels in Eastern Europe, Romania again excepted, took place at a later stage than in the Soviet Union and accelerated during the sixties. However, in the late sixties it was suggested that the Soviet Union was beginning to face problems in the oil and gas industries, which might result in its being unable to meet Comecon requirements, if the rates of growth in consumption recorded in the sixties were to continue.[54] At this time all the Comecon countries negotiated a number of small oil supply contracts with OPEC producers, largely on the basis of barter trade.

The Soviet Union enjoys substantial earnings from trade in oil and gas, and in view of the apparent need to import goods from the developed capitalist world, the marked increase in energy prices has worked to Soviet advantage. However, this has coincided with increasing economic and logistic problems in the Soviet oil and gas industries, necessitating a difficult choice between the opportunities presented by the Western countries' need for energy and the desirability of maintaining the economic cohesion of Comecon, in which hard currency reserves are believed to be limited. The capacity of the Eastern European members to expand bilateral trade with non-Soviet hydrocarbon producers is similarly limited.

The Soviet Plan for 1976-1980 was formulated in the light of shortfalls in the previous Plan in both the oil and gas industries, rising demand in Comecon and a much-increased import price for OPEC oil. Given that the development of Siberia is now regarded as the basis of future Soviet oil and gas development, there are considerable economic and technical problems in raising production. Soviet oil and gas production and transportation is subject to increasing costs. Consequently the role of oil and gas in the domestic energy balance and as export commodities in relation to alternative energy sources has been reassessed.

If Comecon can remain self-sufficient in energy, then the rising world price for oil and gas and the desire of the developed Western countries to diversify their sources of supply work to the advantage of the Soviet Union. Thus OPEC's pricing policy would continue to attract active Soviet support. If on the other hand increased imports are inevitable, then Soviet interests might be served better by attempting to dissuade OPEC from raising prices. Alternatively, if the Soviet Union felt powerless to influence OPEC, then, recognizing the inevitable, it might have to readjust the domestic energy balance in relation to a restricted ability to import, and an enhanced opportunity for export. These issues are complicated by the fact that the principal oil producing area, the Middle East, has been an arena of superpower rivalry long before domestic energy questions could have been considered as influencing the foreign policy of either of the principal protagonists, the Soviet Union and the United States.

An appreciation of the major postwar trends to 1970 is essential for understanding the issues that have influenced policy in the seventies. A number of Western publications deal with this period. The analysis of Demetri B Shimkin[55] traced the development of Soviet fuels production from 1928 to 1958 and provided, in a number of statistical tables, the trends in output, employment and labour productivity. Robert E Ebel's work, entitled *The Petroleum Industry of the Soviet Union*,[56] was essentially a report of a visit to the Soviet Union made in 1960 by a delegation from the US petroleum industry and government officials. It provided particularly detailed information on refining and production, but the concentration was technical rather than economic. The major analytical work of the pre-1970 period was that of Robert W Campbell, published in 1968.[57] This gave a detailed account of the economics and technology of the Soviet oil and gas industries, tracing their development to 1967, and contained an excellent bibliography of Soviet publications dealing with development in the fifties and sixties. The growth, decline and re-emergence of the Soviet Union as an oil trader and the commencement of gas trade is outlined in a further work of Ebel, published in 1970.[58] This work covered the period from the origins of the Russian oil industry in the late nineteenth century to 1967. Chapter 2 of the present study relies on the aforementioned publications.

Iain Elliot's study of the development of Soviet energy,[59] published in 1974, gave a detailed historical exposition of each of the Soviet fuel industries to 1973, and included some data on fuel consumption patterns. A further work of Campbell[60] advanced the analysis of his 1968 work and provided a valuable assessment of the factors which have led certain other Western observers to conclude that the Soviet Union now faces an 'energy crisis'. Western analysis of the energy position in Eastern Europe is somewhat sparse, and confined to a few short articles cited in the text of the present study.

The work of Jochen Bethkenhagen,[61] published in 1975, was a detailed account of growth and prospects in Soviet trade in coal, oil and gas, covering the period from 1960 to 1980. However, the most recent data on which the work was based were for early 1973. Consequently the impact of OPEC's action in 1973-1974 was not discussed. Bethkenhagen's work has nonetheless been a useful basis for chapter 6 of the present study.

There is now a substantial amount of published work on Soviet-Middle East relations, some of which includes analysis of the energy question in relation to Soviet policy in the area. The work of Jeremy Russell,[62] published in 1976, sought to establish whether there was a link between domestic energy problems and the conduct of foreign policy. Valuable though the work is as a general historical account of trade developments, it overlooked the single important factor determining the recent changes in the world market and international energy relations, namely the enhanced power of OPEC and *its* ability to influence not only energy policy but also economic and political relations between the established superpowers.

This dimension is developed to a degree in the work of the American political scientist Arthur Jay Klinghoffer.[63] He outlined the development of oil and gas trade links between the Soviet Union and Middle Eastern producers from the mid-fifties to the immediate aftermath of the Yom Kippur War, stressing the build-up of Soviet interest in the area and the emergence of an independent OPEC which, whilst not by definition a homogeneous politico-military superpower, became an economic superpower at a time when the established adversaries were committed to a policy of detente. These works, along with other publications of the late sixties and early seventies, are examined and reassessed in chapter 7. The Soviet response to developments by non-Middle East hydrocarbon producers, particularly Norway, is also discussed.

Such are the complex and fascinating issues that form the subject matter of this book. In studying the development of the Soviet oil and gas industries one learns much about the growing pains of this industrial giant, facing problems of coordination in a rapidly changing internal and external economic environment with plans on which an entire industrial system depends for a specified period. The problems faced in Comecon hydrocarbon development and the degree of success achieved by Comecon planners, economists and technologists influence not only the nature and extent of economic development in the East, but also its foreign trade structure and role in international affairs.

References

1. These include five American companies: Exxon, based in New York and trading for the most part under the Esso name; Standard Oil of New York (Mobiloil); Gulf Oil, based in Pittsburgh; the Texas Oil Company (Texaco), one of the prime movers in the development of the domestic oil industry but less involved than the above in international trade; and Standard Oil of California, trading under the 'Chevron' name. To these are added the British Petroleum Company (49 per cent state-owned) and the Shell International Petroleum Company (60 per cent Dutch- and 40 per cent British-owned).

2. A recent detailed and objective analysis of the development of the competitive process in the international oil industry is given by Neil H Jacoby, in *Multinational Oil*, New York, Macmillan, 1974, pp 120-149.

3. US policy stipulated that imports of crude oil and petroleum products should be restricted to a maximum of 10 per cent of demand. This figure proved unenforceable and was subject to revision before finally being abolished in mid-1973.

4. A detailed analysis of the decline of the Western European coal industry is given in Richard L Gordon, *The Evolution of Energy Policy in Western Europe: The Reluctant Retreat from Coal*, New York, Praeger, 1970.

5. P R Odell, *The Western European Energy Economy*, Leiden, Stenfert Kroese, 1976, p 18.

6. Jacoby, *op cit*, pp 172-212, gave an analysis of the structural change in the international oil industry to 1972.

7. As did James E Akins, for example, then Adviser on Oil Affairs to the US State Department, at a meeting of OAPEC in 1972. Odell (*op cit*, p 20) made the point that American industry was 'fed up' with the advantage enjoyed by European manufacturers as a result of higher energy costs in the US following the government's decision to maintain import quotas.

8. Odell, *op cit*, p 21.

9. Philip Windsor, *Oil: A Plain Man's Guide to the Energy Crisis*, London, Temple Smith, 1975, p 103.

10. Edith T Penrose, 'The Development of Crisis, *Daedalus*, Vol 104, No 4 (Fall 1975), p 41.

11. For an analysis of the circumstances leading to the oil companies' loss of operating flexibility, see J D Park, 'OPEC and the Superpowers: An Interpretation', *Co-existence*, April 1976, pp 49-64.

12. Resolution XXI.120 of the OPEC meeting, Caracas, December 1970, recorded in Penrose, *op cit*, p 42.

13. *ibid*, p 53.

14. This is discussed by Y Tsumuri in 'Japan', *Daedalus*, Vol 104, No 4 (Fall 1975), pp 125-126.

15. *Petroleum Economist*, June 1974, p 215.

16. Tsumuri, *op cit*, p 113.

17. For a description of Japanese involvement in these areas, see R P Sinha, 'Japan and the Oil Crisis', *The World Today*, August 1974, pp 339-341.

18. *ibid*, p 339.

19. *Petroleum Press Service*, November 1970, p 402.

20. Richard B Mancke, 'Genesis of the US Oil Crisis' in J S Szyliowicz, B E O'Neill (eds), *The Energy Crisis and US Foreign Policy*, New York, Praeger, 1975, p 59.

21. Park, *op cit*, p 56; *Petroleum Economist*, March 1976, pp 84-86.

22. American companies are currently discussing possible imports of gas from Canada, Alaska, Venezuela, Trinidad, Ecuador, Algeria, Nigeria, Australia, Indonesia and East and West Siberia. See Patricia E Starratt, Robert M Spann, 'Alternative Strategies for Dealing with the Natural Gas Shortage in the United States', in E W Erickson, L Waverman (eds), *The Energy Question: An International Failure of Policy*, Toronto, University of Toronto Press, 1974, vol 2, p 39.

23. R El Mallakh, 'American-Arab Relations: Conflict or Cooperation?', *Energy Policy*, September 1975, p 172.

24. F G Zarb, 'US Energy Policy', *The World Today*, January 1976, p 1.

25. P R Odell, 'Europe's Oil', *National Westminster Bank Quarterly Review*, August 1972, p 6.

26. *ibid*, p 12.

27. Derek Ezra, 'Possibilities of a World Energy Crisis', *National Westminster Bank Quarterly Review*, November 1972, p 31.

28. *ibid.*

29. *ibid*, pp 32-33.

30. P R Odell, 'European Alternatives to Oil Imports from OPEC Countries', in F A M Alting von Geusau (ed), *Energy in the European Communities*, Leiden, Sijthoff, 1975, p 65. Note also that prior to the OPEC price rises of 1973-1974, Odell argued the case for European self-sufficiency based on North Sea gas. See 'Indigenous Oil and Gas and Western Europe's Energy Options', *Energy Policy*, June 1973, pp 47-64.

31. Odell, *'European Alternatives . . .'*, (1975), p 71.

32. M A Adelman, 'Is the Oil Shortage Real?', *Foreign Policy*, No 9 (Winter 1972-1973), pp 69-107.

33. Odell, *'European Alternatives . . .'*, (1975), p 71.

34. 'Had the (European Economic) Community confronted the issue, it would have realised that the price paid for increased security would be the increased domestic production of energy – an expensive policy. The United States was already paying this price, but the European coal lobby was too weak to impose it . . .', A Prodi, R Clo, 'Europe', in *Daedalus*, Vol 104, No 4 (Fall 1975), p 105.

35. OECD, *Energy Prospects to 1985*, Paris, OECD, 1974 (2 vols).

36. *ibid*, Vol 1, p 4.

37. The diversity of objectives in the EEC is discussed by G F Ray, A Dean, 'Possible Approaches to a Common European Energy Policy', in Alting von Geusau (ed), *op cit*, pp 167, 169.

38. The background to Western European over-capacity is summarized in *Petroleum Economist*, September 1976, pp 340-342.

39. Z Mikdashi, *The Community of Oil-Exporting Countries: A Study in Governmental Cooperation*, London, Allen and Unwin, 1972, p 69.

40. *ibid.*

41. Resolution 1.1 of the Baghdad Conference of OPEC, held from 10-14 September 1960, recorded in M S Al-Otaiba, *OPEC and the Petroleum Industry*, London, Croom Helm, 1975, p 58.

42. Edith T Penrose, *The Large International Firm in Developing Countries: The International Petroleum Industry*, London, Allen and Unwin, 1968, p 260.

43. '. . . circumstances have changed, and nowhere do governments accept the role of a sleeping partner. They want to have a direct role in the management and the exploitation of national resources, so as to get know-how and to develop national expertise in the production and marketing of oil.' N Al-Pachachi (then Secretary-General of OPEC) in *The Times*, 9 November 1971, p 21.

44. The background to the Libyan price rise is given in Edith T Penrose, 'The Development of Crisis', (1975), pp 41-42. See also *Petroleum Press Service*, January 1971, p 10.

45. *Petroleum Press Service*, March 1971, pp 82-83.

46. *ibid*, May 1971, pp 162-163.

47. Penrose, *op cit*, pp 44, 46.

48. Note that the discussions on the first doubling of prices took place in OAPEC and not OPEC meetings. Likewise discussions on the possible use of an embargo. It was the Arab members who carried through these measures: some producers were hesitant. See Z Mikdashi, 'The OPEC Process', *Daedalus*, vol 104, No 4 (Fall 1975), p 208.

49. P R Odell, 'The World of Oil Power in 1975', *The World Today*, July 1975, p 276.

50. M C Kaser, 'Oil and the Broader Participation of IBEC', *International Currency Review*, 6/1974, p 25.

51. OPEC does encourage the consumer countries to diversify into alternative energy sources (see esp Sheikh Yamani, Saudi Minister of Petroleum, in E J Mitchell (ed), *Dialogue on World Oil*, Washington DC, American Enterprise Institute for Public Policy Research, 1974, p 93). The point is that OPEC as a group now has and uses power to influence decisions on substitution levels.

52. *ibid*, p 99.

53. The Siberian share of total Soviet energy reserves was put at 65 per cent in 1968, A E Probst (ed), *Razvitie toplivnoi bazy raionov SSSR*, Moscow Nedra, 1968, p 38, and at 90 per cent in 1975, editorial to *Planovoe khozyaistvo*, 2/1975, p 5.

54. (i) J G Polach, 'The Development of Energy in East Europe', in *Economic Developments in Countries of Eastern Europe*, Washington DC, US Government, Joint Economic Committee, 1970, pp 348-433; (ii) S Wasowski, 'The Fuel Situation in Eastern Europe', *Soviet Studies*, July 1969, pp 35-51.

55. D B Shimkin, *The Soviet Mineral-Fuels Industries 1928-1958*, Washington DC, US Department of Commerce, 1962.

56. R E Ebel, *The Petroleum Industry of the Soviet Union*, New York, American Petroleum Institute, 1961.

57. R W Campbell, *The Economics of Soviet Oil and Gas*, Baltimore, Johns Hopkins Press, 1968.

58. R E Ebel, *Communist Trade in Oil and Gas*, New York, Praeger, 1970.

59. I S Elliot, *The Soviet Energy Balance*, New York, Praeger, 1974.

60. R W Campbell, *Trends in the Soviet Oil and Gas Industry*, Baltimore, Johns Hopkins, 1976.

61. Jochen Bethkenhagen, *Bedeutung und Möglichkeiten des Ost-West-Handels mit Energierohstoffen* (Deutches Institut für Wirtschaftsforschung, Sonderheft 104), Berlin, Duncker & Humblot, 1975.

62. Jeremy Russell, *Energy as a Factor in Soviet Foreign Policy*, Farnborough, Saxon House, 1976.

63. Arthur Jay Klinghoffer, *The Soviet Union and International Oil Politics*, New York, Columbia University Press, 1977.

Chapter 2

The Postwar Development of the Soviet Oil and Gas Industries to 1970

In examining the development of the Soviet oil and gas industries, it is convenient to divide the years from 1945 to 1970 into two periods. The first dates from 1945 to 1955, in which the oil industry recovered from the severe decline experienced in the Second World War and regained its status as a net exporter. The second dates from 1955 to 1970, and in this period the oil industry expanded steadily, providing the bulk of the Soviet Union's growing energy requirement; it also formed the basis of the Soviet drive to expand the chemical industry, initiated by Khrushchev. It was during this latter period that the potential of the gas industry was realized and that decisions were taken concerning its rapid development. A number of problems arose in the oil industry towards the end of this period which cast some doubt on the future role of oil within the Soviet energy balance, and as a leading export commodity. This obliged Soviet planners to reconsider the relative importance of oil against that of coal and gas, and the possible emergence of natural gas as an export fuel.

Soviet Oil and Gas Industries, 1946-1955

During the Second World War the problems of a declining rate of growth in output of crude oil, experienced since the mid-thirties, were intensified. The Krasnodar oilfields were lost to the advancing Germans and operations in Groznyi, though not entirely lost, were closed down. Between 1941 and 1945 there was a decline in output of crude oil from 33 to 19.4 million tonnes.[1] Through the Lend-Lease system the US came to the aid of the Soviet Union in providing supplies of refined products, primarily aviation gasoline, and these shipments were continued in 1946 and 1947. Table 2.1 shows the pattern of Soviet production and imports from 1942 to 1945.

The concentration of production in Caucasian fields proved to be a liability during the War (the geographical distribution of Soviet oil production is outlined in table 2.2), since the Northern fields were eventually overrun. However, this fact in itself prompted the Soviet Union into undertaking exploratory work in the Volga-Ural area, though it was not until late in 1946 that a substantial discovery, the Tuymazy field, was made.[2]

The potential of the Volga-Ural area had been appreciated during the thirties, and championed by the geologist I M Gubkin.[3] However, even the modest targets set for oil production in this area during the thirties were underfulfilled. In his study of the development of the Soviet oil and gas industries written in 1968,

Table 2.1 Soviet Oil Production and Imports from the US 1941-1945 (million tonnes)

Year	*Production*	*Index*	*Imports*
1941	33.0	100	0.301
1942	22.0	66.7	0.149
1943	18.0	54.5	0.362
1944	18.3	55.5	0.609
1945	19.4	58.8	0.539

Source: R E Ebel, *Communist Trade in Oil and Gas*, New York, Praeger, 1970, pp 28-29

Table 2.2 Regional Distribution of Soviet Oil Production at end of 1940

Region	*Production (million tonnes)*	*% of Total*
Caucasus (Baku and Groznyi)	27.05	87.1
Volga-Ural	1.85	6.0
Kazakhstan and Central Asia	1.50	4.8
Others (Ukraine, North, Far East)	0.70	2.1
Total	31.10	100.0

Source: M M Brenner, *Ekonomika neftyanoi promyshlennosti*, Moscow, Nedra, 1962, p 51

Campbell attributed this delay in appreciation and exploitation of the Volga-Ural fields to disbelief on the part of Soviet central administrators as to the extent of reserves. There was also a lack of confidence in the available drilling technology and, perhaps most important, an inflexibility in the attitude of the planners. In Campbell's view, it took the catalyzing effect of the Second World War to evoke real interest in these fields.[4] During the thirties there may have been some rationale in the planners' policy of non-development of the Volga-Ural fields on technical grounds. For example, it is known that at this time the absence of appropriate technology to maintain wellhead pressures resulted in very low extraction rates for individual wells.[5] It is also recorded that in 1937 only 8,207 out of a total of 12,623 wells drilled in the Soviet Union were operational, the remainder were under repair or awaiting refurbishment.[6] By 1940 output per well had dropped, as had labour productivity.[7] Given this situation, the risk of channelling valuable material and labour resources into the Volga-Ural area was judged as being too high.

The development of the Volga-Ural fields gathered pace in the immediate postwar period, facilitated by the availability of improved and in some cases imported technology. By 1955, the requirement in number of wells in a given unit area had been reduced to one-sixth of the prewar level[8] and depths of 1,400 metres were attained in production drilling, giving access to the major oil-bearing layers of this area. As Campbell indicated,[9] the nature of Volga-Ural crude oil posed considerable problems in the refining industry: the majority of the oils were found to be heavy, of high sulphur and paraffin wax content, and the gasoline fractions derived therefrom tended to be of low octane rating. Thus further processing was needed to bring them to the required standard. According to a Soviet source of the late fifties the capital costs for oil refining at that time rose markedly with the sulphur content of the oil. Taking the index of the refining cost of oil with a sulphur content of less than 0.5 per cent as 100, the corresponding index for oil with a sulphur content of 0.6 to 1.9 per cent was 150 to 200, and that for oil with a sulphur content of 2.0 and over, 220 to 300.[10] Despite the fact that the majority of Volga-Ural crude oils belonged to the last category, the economics of scale in production eventually outweighed the additional refining costs. This was especially the case after the development of the Romashkino field in the Tatar republic, discovered in 1948 and brought fully on stream in the early fifties, at which time it was thought to be the world's largest single field.

Table 2.3 Soviet Oil Production by Region 1945-1955 (thousand tonnes)

	1945	*1950*	*1955*
Soviet Total	19,436	37,878	70,793
RSFSR	5,675	18,231	49,263
Volga-Ural	2,833	10,990	41,220
Tatar ASSR	10	1,020	14,600
Bashkir ASSR	1,300	5,250	14,200
Kuybyshev oblast'	1,020	3,480	7,250
Stalingrad oblast'	na	na	2,300
Saratov oblast'	na	440	1,900
Orenburg oblast'	na	na	510
Perm oblast'	190	300	570
West Siberia	–	–	–
North Caucasus	na	6,310	6,540
Krasnodar krai	700	3,000	3,890
Chechen-Ingush ASSR	890	2,500	2,120
Dagestan ASSR	550	920	520
Stavropol krai	na	–	neg
Sakhalin	na	620	950
Komi ASSR	na	330	550
Belorussia	–	–	–
Ukraine	250	293	531
Turkmenia	629	2,021	3,162
Fergana Valley (Uzbekistan)	517	1,409	1,128
Georgia	36	43	43
Kazakhstan	788	1,059	1,397

(– = zero, neg = negligible, na = not available)

Source: R W Campbell, *The Economics of Soviet Oil and Gas*, Baltimore, Johns Hopkins Press, 1968, p 124

The share of Soviet production held by the Volga-Ural region rose to 29 per cent in 1950 and 58 per cent in 1955,[11] accounted for by the output and growth of the Romashkino field. Another major area for expansion in oil production between 1946 and 1955 was the Bashkir ASSR. And by 1955 the

Azerbaidzhan republic had also gone some way towards regaining its prewar peak production level of 21.4 million tonnes, recorded in 1937. The pattern of Soviet oil production from 1945 to 1955, by region, is detailed in table 2.3.

By world standards refinery technology was at a low level and this gave rise to a pattern of product output and final consumption which had to be modified in the latter part of the eighth Five-Year Plan. During the fifties Soviet refineries were designed and geared to maximal output of fuel oil and as a result the initial conversion of the Soviet rail network from coal was to fuel oil rather than to diesel oil. An additional factor was the quality of Soviet motor fuel which was particularly low, ranging from 50 to 70 octane.

Writing in 1962, the American analyst Demetri Shimkin noted that the structure of Soviet fuel output had changed very little between 1928 and 1958. He expressed the view that such a rigid policy placed considerable strain on the economy as a whole, but particularly on the transport sector, preventing substantial savings in conversion from steam to internal combustion engines, and from large wood-fired to oil-fired furnaces. This occurred despite cost-price relations that strongly favoured the use of oil and gas in preference to coal and other solid fuels.[12]

As Alec Nove has indicated,[13] the system of planning by material balances was based inevitably on past experience. It ignored the often substantial effect of change in the period immediately preceding the Plan legislation and the need to make adjustments in the light of further changed circumstances. Specifically in the case of energy planning, the Stalinist system which applied in the postwar period operated on the basis of the extrapolation of a simple energy to NMP coefficient, the result being translated into standard fuel units and subsequently into physical units in accordance with unchanging relativities between fuels. Not only was there a considerable economic loss in the fuel economy, but there was also a delay in the development of the Soviet chemical and petrochemical industries. Irrational fuel prices served further to complicate the issues. Shimkin's analysis points to the fact that in 1955 delivered prices for hard and brown coal covered only 71 and 91 per cent of costs respectively, whereas the average selling price of petroleum products was some 90 per cent above cost and included substantial price discrimination between consumers.[14] Shimkin did, however, note that the 'uneconomic outlook' of planners became modified after 1950.[15]

On the question of the difficulties encountered in determining an optimal energy balance, Campbell's view was that the planning system 'has erected too many partitions *within* the problem, and hampered joint consideration of interrelated problems'.[16] Though plans for economic growth during the period of Soviet reconstruction were broadly fulfilled in the quantitative sense, the energy economy was far from healthy. It was characterized by relatively inefficient conversion processes, reflected in high levels of energy consumption compared with final output.

Soviet Oil and Gas Production, 1956-1970

By the time of the sixth Five-Year Plan it was evident that considerable rethinking had taken place concerning the role of oil and gas within the Soviet energy economy. In an article published in March 1956, the Soviet analysts Agukin and Shakhmatov indicated the need to expand the use of oil and gas in preference to other energy sources.[17] They stated that in 1955 the average production cost of one tonne of standard fuel in the form of oil was less than one quarter of that of coal, one eighth of shale and one twelfth of peat. Production of one tonne of naphtha from oil was almost three times cheaper than by the liquefaction and hydrogenation of coal.[18] Gas was stated to have even greater advantages, the average production cost, on a standard fuel basis, was one twelfth of that of Moscow coal and shale, one seventh of peat and less than half that of oil.[19]

Despite these advantages, which it is reasonable to suppose must have been appreciated earlier, the gas industry was comparatively underdeveloped. Agukin and Shakhmatov pointed to inadequate prospecting and delays in the construction of production centers when useful deposits were found.[20] Accordingly the directives of the sixth Plan reflected a sharp rise in the production of gas, envisaging a 3.9-fold increase on the 1955 production level, reaching 35 billion cubic metres in 1960. At the same time the target for discovery of new reserves in the A+B+C1 categories represented an increase of 85 to 90 per cent for the five-year period.[21]

It is not within the scope of this volume to explore the reasons for the abandonment of the sixth Plan or for the non-acceptance of the seventh: suffice it to note, as Nove indicated,[22] that a major reason given for the preparation of a Plan covering the seven-year period from 1959 to 1965 was the discovery of new fuel reserves and Khrushchev's desire to develop rapidly the chemical and petrochemical industries. A point of interest is that the gas production figure for 1958 was 29.9 billion cubic metres and that it rose to 45.3 billion in 1960, this latter figure was well in excess of the level envisaged by Agukin and Shakhmatov when the sixth Plan was current.[23]

The discovery and development of the Volga-Ural oilfields prior to the Seven-Year Plan facilitated the rapid growth of oil production in the Soviet Union to 1965. Campbell noted that the earliest fields to be exploited in the Volga-Ural area were comparatively large and the concentration on exploration meant that by 1959 some 80 per cent of Soviet oil reserves in the A+B categories were reckoned to be located in this area.[24] In addition the oilfields were located close to one another, enabling a rapid rate of reserve accumulation to be attained at relatively low cost. The initial output per well proved to be higher than the national average, thus minimizing the cost of production drilling.[25]

There was also a shift in the regional pattern of gas production. The major part of the 9 billion cubic metres of gas produced in the Soviet Union in 1955 came from fields located in European Russia and the Ukraine. During the Seven-Year Plan the priority areas were the Volga fields, the North Caucasus, the Ukraine and Central Asia.[26] Iain Elliot in his study of the historical development of the Soviet energy balance written in 1974, pointed out that the storage of gas is considerably more complicated and more expensive than that of other conventional fuels, and early Soviet production policy provided for the exploitation of fields located close to potential consumers.[27] This was the case even when the reserves of individual deposits were comparatively limited. The benefit of rapid natural gas development during the sixties is illustrated by the fact that the coal and oil industries provided a return of 48 tonnes of standard fuel per thousand rubles invested, compared with 296 tonnes in the gas industry. Output per worker in the coal industry in the late sixties was 38 tonnes of standard fuel annually compared with 330 tonnes in the oil industry and 2,100 tonnes in the gas industry.[28]

During the Seven-Year Plan the oil industry fulfilled its objectives, recording a production level of 243 million tonnes in 1965 against the Plan projection of 230-240 million. However, the gas industry reached a level of only 129 billion cubic metres against the Plan projection of 150. This tightening in the supply of hydrocarbon energy was exacerbated by the underfulfilment of the Plan by the coal industry, necessitating periodic revisions of annual targets.[29] The situation was further compounded by a declining rate of growth in output of the major fuels during the eighth Plan, when the average annual rate was approximately 5.5 per cent compared with 7.3 per cent between 1961 and 1965. Rising demand for energy and depletion of conveniently located reserves – resulting in the need to develop more remote deposits, especially of oil and gas – caused stagnation in oil exports and gave rise to the commencement of gas imports. In 1969, the lowest annual rate of increase in oil production since the Second World War, 6.1 per cent, was recorded: this was, however, rectified in 1970 when the rate was 7.6 per cent, bringing total production to 353 million tonnes. In 1969 gas production totalled 181 billion cubic metres, a shortfall against the Plan of 3 billion, but constituting a 7.2 per cent increase on the previous year. As a result the initial target for 1970 of 200 billion cubic metres was lowered to 196, and this was 'fulfilled'. The final figure was 198 billion, a 9.4 per cent increase on 1969.[30]

By the start of the eighth Plan the importance of Siberian reserves of oil and gas for the development of the Soviet energy economy was appreciated, though the inevitability of their exploitation had been accepted some 10 years earlier.[31] The 23rd Party Congress held in 1966 decreed that a major economic complex would eventually be created in Siberia on the basis of oil, gas and timber resources, and that the effect of the development of oil and gas would be felt by the end of the ninth Plan.[32]

Table 2.4a details the growth of fuel production in the Soviet Union between 1955 and 1970, showing the steady increase in production of each of the major fuels and the expanding share of the production balance held by oil and gas, at the expense of other primary energy sources. Writing in 1970, the American analyst J R Lee pointed to the decline in production of coal in 1961 compared with 1960, and in 1968 compared with 1967.[33] However, he omitted to mention that the shortfall was recorded only in the production of brown coal, the calorific value of which is roughly half that of hard coal. Production of hard coal, in fact, rose throughout the sixties. Soviet statistics of the changing production balance, expressed in units of standard fuel, present a more accurate picture which can be seen in table 2.4b.

However, this is not to deny that the Soviet coal industry faced a number of difficulties during the eighth Plan. These included considerable delays in the commissioning of new productive capacity. Lee indicated that in 1968 only 12 million tonnes of new capacity was provided, compared with an average of 17 million tonnes per year between 1961 and 1967. He added the proviso that

Table 2.4a Soviet Fuels Production 1955-1970

Fuel	*1955*	*1956*	*1957*	*1958*	*1959*	*1960*	*1961*	*1962*	*1963*	*1964*	*1965*	*1966*	*1967*	*1968*	*1969*	*1970*
Coal (million tonnes)	391.3	429.2	463.5	496.1	506.6	509.6	506.4	517.4	531.7	554.0	577.7	585.6	595.2	594.2	607.8	624.1
including Hard	276.6	304.0	328.5	353.0	365.2	374.9	377.0	386.4	395.1	408.9	427.9	439.2	451.4	455.9	467.3	476.4
Brown	114.7	125.2	135.0	143.1	141.4	134.7	129.4	131.0	136.6	145.1	149.8	146.4	143.8	138.3	140.5	147.7
Oil (million tonnes)	70.8	83.8	98.3	113.2	129.6	147.8	166.1	186.2	206.1	223.6	242.9	265.1	288.1	309.2	328.4	353.0
Gas (billion cubic metres)	9.0	12.1	18.6	28.1	35.4	45.3	59.0	73.5	89.8	108.6	127.7	143.0	157.4	169.1	181.1	197.9
% Gain over Previous Year		*1956*	*1957*	*1958*	*1959*	*1960*	*1961*	*1962*	*1963*	*1964*	*1965*	*1966*	*1967*	*1968*	*1969*	*1970*
Coal		9.7	8.0	7.0	2.1	1.3	−0.6	2.2	2.8	4.2	4.3	1.4	1.6	−0.2	2.4	2.7
including Hard		9.9	8.1	7.5	3.5	2.7	0.6	2.5	2.2	3.5	4.6	2.6	2.8	1.0	2.5	1.9
Brown		9.2	7.8	5.9	−1.2	−4.7	−3.9	1.2	4.3	6.2	3.2	−2.3	−1.8	−3.8	1.2	5.1
Oil		18.3	17.3	15.2	14.5	14.1	12.3	12.1	10.6	8.5	8.6	9.2	8.7	7.3	6.1	7.5
Gas		34.4	53.7	51.1	26.0	28.0	30.2	24.7	22.2	20.9	17.6	12.0	10.1	7.4	7.2	9.3

Sources: *Narodnoe khozyaistvo SSSR, 1958*, p 213; *1960*, pp 257, 262, 267; *1965*, pp 175, 177-178; *1970*, pp 184-185, 187; *1975*, pp 240-242.
J R Lee, 'The Fuels Industries', in *Economic Performance and the Military Burden in the Soviet Union*, Washington DC, US Congress, Joint Economic Committee 1970, p 36

Table 2.4b Soviet Fuels Production Balance 1955-1970 (million tonnes standard fuel)

Year	*Total*	*Oil (including gas condensate)*	*Gas*	*Coal*	*Peat*	*Shale*	*Wood*
1955	479.9	101.2	11.4	310.8	20.8	3.3	32.4
1956	514.0	119.8	15.2	325.1	18.4	3.5	32.0
1957	574.6	140.6	23.2	351.7	22.5	3.7	32.9
1958	616.4	161.9	33.9	362.1	21.1	4.5	32.9
1959	659.4	185.3	42.5	370.0	23.0	4.6	34.0
1960	692.8	211.4	54.4	373.1	20.4	4.8	28.7
1961	732.7	237.5	70.8	370.7	19.5	5.2	29.6
1962	778.6	266.5	84.6	379.7	12.9	5.8	29.1
1963	847.1	294.7	105.1	388.4	21.7	6.5	30.7
1964	912.2	319.8	127.0	403.3	22.2	7.1	32.8
1965	966.6	346.4	149.8	412.5	17.0	7.4	33.5
1966	1,033.1	379.1	170.1	420.1	24.4	7.5	31.9
1967	1,088.4	411.9	187.4	428.6	22.4	7.5	30.6
1968	1,126.6	442.1	201.2	428.7	18.3	7.6	28.7
1969	1,177.4	469.6	215.5	439.6	16.7	8.0	28.0
1970	1,221.8	502.5	233.5	432.7	17.7	8.8	26.6

Sources: 1956-1959, *Narodnoe khozyaistvo SSSR, 1960*, p 253
1961-1964, *Narodnoe khozyaistvo SSSR, 1972*, p 209
1955, 1960 and 1965-1970, *Narodnoe khozyaistvo SSSR, 1975*, p 239

Table 2.4c Soviet Fuels Production Balance 1955-1970 (%)

Year	*Total*	*Oil (including gas condensate)*	*Gas*	*Coal*	*Peat*	*Shale*	*Wood*
1955	100	21.1	2.4	64.8	4.3	0.7	6.7
1956	100	23.3	3.0	63.2	3.6	0.7	6.2
1957	100	24.5	4.0	61.2	3.9	0.7	5.7
1958	100	26.3	5.5	58.8	3.4	0.7	5.3
1959	100	28.1	6.4	56.1	3.5	0.7	5.2
1960	100	30.5	7.9	53.9	2.9	0.7	4.1
1961	100	32.4	9.7	50.5	2.7	0.7	4.0
1962	100	34.2	10.9	48.8	1.7	0.7	3.7
1963	100	34.8	12.4	45.9	2.5	0.8	3.6
1964	100	35.1	13.9	44.2	2.4	0.8	3.6
1965	100	35.8	15.5	42.7	1.7	0.8	3.5
1966	100	36.7	16.5	40.7	2.3	0.7	3.1
1967	100	37.8	17.2	39.4	2.1	0.7	2.8
1968	100	39.2	17.9	38.0	1.6	0.7	2.6
1969	100	39.9	18.3	37.3	1.4	0.7	2.4
1970	100	41.1	19.1	35.4	1.5	0.7	2.2

Sources: 1956-1959, *Narodnoe khozyaistvo SSSR, 1960*, p 253
1961-1964, *Narodnoe khozyaistvo SSSR, 1972*, p 205
1955, 1960 and 1965-1970, *Narodnoe khozyaistvo SSSR, 1975*, p 239

refurbishment of the industry in the latter half of the eighth Plan should render the 1970 production target of 618 million tonnes readily attainable.[34] In the event this target was exceeded by just over 6 million tonnes.

In 1968 oil replaced coal as the major fuel produced in the Soviet Union, its share of the production balance having risen from 21.1 per cent in 1955 to 39.2 per cent. In 1970 this share had risen to 41.1 per cent. Between 1955 and 1970 the share of gas in the production balance rose from 2.4 to 19.1 per cent. Oil production in individual republics is detailed in table 2.5, showing the predominance of the RSFSR, which includes the Volga-Ural fields and the emerging West Siberian deposits, which began to contribute in the latter years of the eighth Plan. The trend in the older areas, Uzbekistan, Azerbaidzhan and Turkmenia, shows a plateau in production after 1965, this being counterbalanced by steady growth in the Ukraine, Belorussia, and Kazakhstan.

The corresponding pattern of gas production is given in table 2.6 and shows continuous growth in the Krasnodar and Stavropol krais and the Volgograd and Saratov oblasts, supplemented by the contribution of the Tyumen' oblast' after 1967. However, as Elliot stated,[35] it was appreciated that in all but the Tyumen' and Volgograd oblasts the rate of discovery of new reserves under the eighth Plan did not keep pace with the increase in production, indicating the likelihood of falling production after 1970, in the areas so affected.

Table 2.7 details the development of the Soviet energy balance to 1970, reflecting the impact of changes in the production of major fuels on consumption, and illustrating the emergence of hydrocarbon energy after 1950. The rates of development of the oil and gas industries gave planners cause for confidence. Their view was that despite the problems encountered towards the end of the eighth Plan, oil and gas would be the prime contributors to growing energy demand in the following decade, potentially accounting for over 60 per cent of the Soviet fuel and energy balance by 1980.[36]

The Development of Oil and Gas Trade

Robert E Ebel's *Communist Trade in Oil and Gas*, published in 1970, provided a detailed history of the Soviet Union as an exporter and importer of oil and gas from the pre-revolutionary origins of the oil industry, to the development of the oil and gas industries in the sixties. It contained statistical information up to and including 1968, and the material in this section is derived to a great extent from this work.

As referred to earlier, imports of certain oil products were essential during the Second World War, as part of the Soviet oil industry fell into enemy hands. Exports recommenced in 1946 but these were counterbalanced by continuing imports of oil and products, such that trade served primarily to optimize the refining balance within Comecon.

The Soviet Union imported kerosene and middle distillate fuels from Romania and East Germany. In 1954 the role of foreign trade was re-evaluated in the process of de-Stalinization and in the gradual easing of the Cocom embargo, exports of oil and products to Western countries were resumed in 1951.[37] These were directed predominantly to Italy and Finland.

Table 2.5 Soviet Oil Production by Republic 1955, 1960 and 1965-1970
(thousand tonnes, including gas condensate)

	1955	*1960*	*1965*	*1966*	*1967*	*1968*	*1969*	*1970*
USSR Total	70,793	147,859	242,888	265,125	288,068	309,150	328,299	352,574
RSFSR	49,263	118,861	199,929	217,982	234,950	251,545	265,653	284,753
Ukraine	531	2,159	7,580	9,288	10,969	12,130	13,351	13,909
Belorussia	–	–	39	208	817	1,718	2,760	4,234
Uzbekistan	996	1,603	1,800	1,721	1,755	1,848	1,799	1,805
Kazakhstan	1,397	1,610	2,022	3,103	5,602	7,429	10,124	13,161
Georgia	43	34	30	30	32	30	26	24
Azerbaidzhan	15,305	17,833	21,500	21,729	21,605	21,138	20,420	20,187
Kirghizia	115	464	305	311	310	306	286	298
Tadzhikistan	17	17	47	81	104	127	155	181
Turkmenia	3,126	5,278	9,636	10,672	11,924	12,879	13,725	14,487

Sources: *Narodnoe khozyaistvo SSSR, 1956*, p 75; *1960*, p 264; *1965*, p 175; *1968*, p 234; *1970*, p 184

Table 2.6 Soviet Natural Gas Production by Republic 1955, 1960 and 1965-1970
(million cubic metres)

	1955	*1960*	*1965*	*1966*	*1967*	*1968*	*1969*	*1970*
USSR Total	8,981	45,303	127,666	142,962	157,445	169,101	181,121	197,945
RSFSR	4,291	24,412	64,257	69,042	74,781	78,347	80,993	83,321
Ukraine	2,928	14,286	39,362	43,617	47,443	50,942	55,403	60,887
Belorussia	–	–	–	–	–	–	–	178
Uzbekistan	103	447	16,474	22,566	26,638	28,988	30,769	32,094
Kazakhstan	25	39	29	46	83	321	680	2,092
Azerbaidzhan	1,494	5,841	6,180	6,173	5,771	4,993	4,938	5,521
Moldavia	–	2	–	–	–	–	–	–
Kirghizia	–	41	155	163	256	291	341	367
Tadzhikistan	–	–	52	90	245	366	438	388
Turkmenia	141	234	1,157	1,265	2,226	4,843	7,535	13,107

Sources: *Narodnoe khozyaistvo SSSR, 1958*, p 213; *1960*, p 267; *1963*, p 156; *1965*, p 177; *1968*, p 236; *1970*, p 185

Table 2.7 The Soviet Energy Balance 1913-1970 (%)

Energy Source	*1913*	*1950*	*1955*	*1960*	*1965*	*1970*
Coal	45.0	64.3	62.5	52.1	41.5	35.0
Oil	24.2	17.0	21.0	29.5	34.7	41.0
Natural Gas	–	2.3	2.3	7.6	15.3	19.0
Other	30.8	16.4	14.2	10.8	8.5	5.0

Sources: I F Elliot, *The Soviet Energy Balance*, New York, Praeger, 1974, p 7
P Hanson, 'The Soviet Energy Balance', *Nature*, Vol 261, May 1976, p 3

By 1954 Sweden had become an importer of Soviet oil under terms which were to become characteristic of the Soviet Union's commercial strategy during the sixties. Sweden was one of the first countries to appreciate its potential vulnerability as a result of a high level of dependence on Middle East oil. As Peter Odell pointed out,[38] Sweden faced a shortlived supply crisis in 1951 as a result of production shortfalls in Iran during that country's nationalization of part of the assets of the British Petroleum Company. Sweden was able to offer what the Soviet Union needed at that time, namely large-diameter steel pipe to develop the oil delivery system. Unencumbered by membership of NATO, which disapproved of the sale by any of its members of steel pipe to the Comecon bloc, and impervious to pressure from neighbouring European states, Sweden built up her barter trade with the Soviet Union to the extent of becoming dependent for some 15 per cent of oil supplies by the mid-fifties.

The Soviet oil trade balance between 1946 and 1955 is detailed in table 2.8. It shows the emergence of the Soviet Union as a net exporter, though still importing some refined products to compensate for the technological backwardness of domestic refineries in relation to the changing pattern of demand.

In the latter half of the fifties the Soviet Union was able to compound her initial success in export trade with a reorientation towards an optimal refining balance, reducing the need to import refined products. As the Volga-Ural fields were brought into production, the need to import crude oil declined. The revival of Comecon coincided with the Soviet Union's resumption of oil exports. Trade within the bloc was re-evaluated, to the extent of cancelling Polish indebtedness on certain credits and by extending new credit facilities to each of the East European countries. The Soviet Union also consented to the termination of postwar reparations from East Germany in 1954, and agreed in principle to base future intra-Comecon trade in all commodities on prices related to those in world markets.[39] Henceforth, as demand for oil grew in Eastern Europe the Soviet Union became the major supplier to all countries except Romania, which was also a net exporter of oil and refined products.

The period from 1956 to 1965 saw the expansion of Soviet oil deliveries to the West at a time when, after the reopening of the Suez Canal and the restoration of production in the Middle East to previous levels, there arose a substantial surplus of oil on the open market. This resulted in the commencement of a downward trend in oil and refined product prices in Western Europe which, coupled with the increasing presence of Soviet oil, caused some measure of concern.

Table 2.8 Soviet Oil Trade Balance 1946-1955 (thousand tonnes)

		1946	*1947*	*1948*	*1949*	*1950*	*1951*	*1952*	*1953*	*1954*	*1955*
Exports:	Oil	0	0	0	100	300	900	1,300	1,500	2,100	2,900
	Products	500	800	700	800	800	1,600	1,800	2,700	4,400	5,100
	Total	500	800	700	900	1,100	2,500	3,100	4,200	6,500	8,000
Imports:	Oil	9.1	74.9	74.0	131.9	336.6	59.9	197.6	104.6	193.0	574.8
	Products	900	500	800	1,700	2,300	2,600	3,600	4,600	3,800	3,800
	Total	909.1	574.9	874.0	1,831.9	2,636.6	2,659.0	3,797.6	4,704.6	3,993.0	4,374.8
Balance	(net export + net import –)										
	Oil	–9.1	–74.9	–74.0	–31.9	–36.6	+841.0	+1,102.4	+1,395.4	+1,907.0	+2,325.2
	Products	–400.0	+300.0	–100.0	–900.0	–1,500.0	–1,000.0	–1,800.0	–1,900.0	+600.0	+1,300.0
	Total	–409.1	+225.1	–174.0	–931.9	–1,536.6	–159.0	–679.6	–504.6	+2,507.0	+3,625.2

Source: R E Ebel, *Communist Trade in Oil and Gas*, New York, Praeger, 1970, p 32

Italy was in the forefront of negotiations with the Soviet Union. The Italian government discovered that by allowing the state oil company ENI to negotiate freely for the supply of crude oil, Soviet oil could be obtained at a delivered cost well below that of Middle East oil supplied through the production and distribution system of the major oil companies. The Soviet Union was prepared in principle to offer an outlet for Italian manufactured goods, and negotiations eventually led to a large-scale bilateral agreement in 1963. This provided for the delivery of 25 million tonnes of Soviet oil over a five-year period in return for steel pipe and later the materials and know-how to build an automobile factory.

However, no export agreement concluded at this time was as dramatic as that negotiated with Cuba shortly after Castro's assumption of power in 1959. At that time Cuba's oil and petroleum products were supplied by three of the major oil companies from their operations in Venezuela. Castro queried the high delivered cost of $3 per barrel in relation to the f.o.b. cost in Venezuela of $2. He then gained Khrushchev's agreement to supply one third of Cuba's requirement at a delivered cost of $2.10 per barrel. When the oil companies (Shell, Esso and Texaco) exercised their legal right to refuse to refine and distribute any oil other than that produced by their own affiliates, Castro succeeded in negotiating with Khrushchev for 100 per cent Soviet supply. Following a directive from Castro to the refineries to process Soviet oil, the companies withdrew their staff and embargoed deliveries to Cuba.[40] The reasons why the companies chose to act in this way are obscure. Perhaps they were acting in alliance with the American government in order to bring pressure to bear on Cuba, whose path of socio-economic development could scarcely have been regarded as attractive. On the other hand the companies might well have underestimated the capacity of the Soviet Union to supply Cuba's entire requirements. Whatever the reason, the companies and the American government lost in the outcome.

On a different plane the emergence of Soviet oil in the world market necessitated a defensive commercial strategy on the part of the major oil companies. The case of the Soviet offer to India is a good illustration of this. By 1960, the end of India's second Five-Year Plan, the balance of payments and the international value of its currency were giving cause for concern. Limited success had been achieved by the Indian government in persuading the oil companies, operating in India, to import crude oil rather than higher-cost refined products. However, rising demand threatened a further drain on an already hard-pressed currency. The Indian government could have chosen to suppress demand, invest in the development of limited domestic hydrocarbon resources or seek recourse to further international loans. It decided against these alternatives and instead sought either to import crude oil and refined products on a barter basis, or to negotiate a price reduction for the oil delivered by the major companies to their affiliates. In mid-1960 the Soviet Union offered crude oil to India at a price some 25 per cent below that being charged by the oil companies. The three companies operating in India lowered their price to counteract the Soviet bid. Since the Soviet Union had just assumed responsibility for supplying Cuba, the companies wished to avoid the consequences of yielding further to market pressure. In addition, the companies convinced the Indian government that they could influence the World Bank and other financial institutions to view favourably subsequent approaches by India for aid.[41]

The presence of the Soviet Union in the world petroleum market caused some consternation in American circles, culminating in the publication of a somewhat alarmist report in 1962 on the so-called Soviet 'oil export offensive'.[42] The report concluded that the Soviet oil export drive had five objectives: protection of the domestic economy against foreign competition, isolation of the currency from the influence of foreign exchange markets, provision of a strengthened bargaining position in relation to capitalist countries, development of the ability to discriminate amongst purchasers and to sell in foreign markets without regard for normal commercial considerations or internal costs, and, lastly, securing flexibility to adjust trade to serve political interests.[43] The problematic issues were seen to arise as a result of the last two objectives. Ebel indicated that Soviet price-cutting was a feature of strategy, but only to the extent necessary to secure a contract, and that every opportunity was taken to maximize revenue from existing business.[44] On the much-publicized question of discriminatory pricing against Eastern Europe, which during the fifties and early sixties was paying more in unit ruble values for Soviet oil and refined products than Western buyers, Ebel rightly drew attention to the fact that as the prices of all commodities in intra-Comecon trade were being brought into line with those of the world markets, the declining price of oil was reflected in the prices charged during the sixties. This was the time when the relative prices of many other goods in intra-Comecon trade were also being re-evaluated.[45] As for the issue of pricing without regard to cost, most Western companies have also at times priced in relation to marginal cost in order to secure export business. In this sense, as Peter Wiles pointed out,[46] the Soviet Foreign Trade Ministry behaves no differently from the export department of a multi-divisional company.

This goes some way towards putting into perspective the 'political' aspect of Soviet oil trade. Political relations are governed by mutual perceptions of policy. If, for example, the US believed that the Soviet offer of oil to Cuba was made for the purpose of gaining a political foothold, then American policy towards the Soviet Union might be influenced by this perception, overlooking the fact that the commercial policy of the oil companies at that time left Castro no alternative. In the Italian case, the Soviet Union had a commercial advantage over the established traders by virtue of the different financial structure of its oil industry, plus the ability to absorb Italian goods as part payment. The subsequent development of this trade has been limited not so much by political considerations but by the capacity of the Soviet Union to absorb what Italian industry was prepared to offer in barter payment. The outcome is a function of the relative competitiveness of bids in the market. The lifting of the NATO embargo in 1966 on deliveries of steel pipe to the Comecon bloc, enabled the Soviet Union to compete more effectively for a share of the West German petroleum market.

In the late sixties there was a fall in deliveries of Soviet oil to non-Comecon markets, which gave rise to the view that the Soviet Union's oil export drive was losing momentum. Much of the slowdown in the growth of net exports was due to a natural delay before the effect of production shortfalls in oil and gas prompted planners to adjust the rate and pattern of growth in domestic demand.

It is likely that Soviet oil export policy was constrained in the late sixties by rising costs of exploration and production, growing demand both domestically and within Eastern Europe, and falling prices in the world market due to intense competition between the oil companies. The fear of Western Europe's possible engulfment in a sea of Soviet oil, expressed in the American National Petroleum Council's 1962 report, was not realized and the future trend of Soviet oil exports appeared uncertain.

Table 2.9 details the Soviet trade balance in crude oil and refined products from 1956 to 1970. It shows the declining import requirement as the Soviet oil production and refining industries expanded, the peak of exports as a percentage of production reached in 1966, and the subsequent decline to 1970.

Table 2.9 Soviet Oil Trade Balance 1956-1970 (million tonnes, Oil and Refined Products)

Year	*Production**	*Exports*	*Imports*	*Net Exports*	*Net Exports as % of Production*
1956	83.8	10.6	5.6	5.0	6.0
1957	98.3	16.3	4.5	11.8	12.0
1958	113.2	18.9	4.6	14.3	12.6
1959	129.6	26.5	4.7	21.8	16.8
1960	147.8	34.5	4.7	29.8	20.2
1961	166.1	42.7	3.8	38.9	23.4
1962	186.2	47.1	3.0	44.0	23.6
1963	206.1	53.1	3.0	50.1	24.3
1964	223.6	58.3	2.3	56.0	25.0
1965	242.9	66.2	2.1	64.1	26.4
1966	265.1	75.6	1.8	73.8	27.8
1967	288.1	81.0	1.5	79.5	27.6
1968	309.2	86.2	1.1	85.1	27.5
1969	328.4	90.8	1.1	89.7	27.3
1970	353.0	95.8	1.1	94.7	26.8

*(*including gas condensate)*

Sources: 1956-1967, R E Ebel, *Communist Trade in Oil and Gas*, New York, Praeger, 1970, p 40; 1968-1970, *Narodnoe khozyaistvo SSSR, 1975*, p 240; *Vneshnyaya torgovlya SSSR, 1969*, pp 25, 38; *1970*, pp 26, 39

Despite the consistent failure, since 1965, of the Soviet gas industry to fulfil Plan targets the Soviet Union has taken advantage of the availability of gas to enter the expanding market in Eastern Europe. The factors determining the growth of the Western European gas market were the development of the Groningen gasfield in the Netherlands, which supplied the domestic market plus West Germany and France; the expansion of Algerian and Libyan deposits, shipped as liquefied petroleum gas to France and Italy respectively; and the development of the Norwegian and British gas reserves in the North Sea. The Norwegian sector provided fuel and feedstock with the possibility of export, given Norway's excess of hydrocarbons over demand; production from the British sector was consumed almost entirely domestically.

The first agreement between the Soviet Union and a Western country for the sale of natural gas was signed with Austria in 1968. In return for Austrian steel plate, which was rolled into pipe in West Germany, the Soviet Union extended the 'Bratstvo' trunk pipeline over the Czech border into Austria and deliveries of Soviet gas commenced at the end of 1968.[47] Other export outlets for Soviet gas were Poland and Czechoslovakia. Small quantities of liquefied gas had been delivered to Poland since 1950,[48] but a substantial volume was supplied after the completion of the northern branch of the 'Bratstvo' pipeline to Warsaw in 1966. Soviet gas supply to Czechoslovakia, which commenced in 1967, helped to alleviate the strains imposed by a relatively rapid rate of economic development and a poor indigenous resource base, which had been supplemented by supplies of Soviet oil. In the late sixties the supply of Soviet oil for export was becoming increasingly problematic. At this time the Soviet Union opened negotiations with Italy, West Germany and Japan concerning the possibility of a future supply of gas in return for material and financial assistance in developing Siberian resources.[49] There were, however, no deliveries of Soviet gas to these countries prior to 1970.

The Soviet Union undertook to import gas from Iran and Afghanistan,[50] but at the same time concluded gas export contracts in Western and Eastern Europe. Soviet imports from these two producers constituted repayment for development aid. Imports from Afghanistan commenced in 1967 and from Iran in 1970. The gas trade balance from 1955 to 1970 is detailed in table 2.10, showing the movement towards a net import position during the late sixties. Despite this trend the medium-term potential for domestic and export substitution of gas for coal was not in doubt. Ebel stated that imports beyond the quantities stipulated in the original agreements with Iran and Afghanistan were not foreseen, and it was expected that in the course of the ninth Plan the Soviet Union would again become a net exporter.[51]

Table 2.10 Soviet Gas Trade 1955-1970 (billion cubic metres)

Year	*Production*	*Exports*	*Imports*	*Net Exports*
1955	9.0	0.139	–	0.139
1960	45.3	0.242	–	0.242
1965	127.7	0.392	–	0.392
1966	143.0	0.828	–	0.828
1967	157.4	1.290	0.207	1.083
1968	169.1	1.729	1.500	0.229
1969	181.1	2.664	2.030	0.634
1970	197.9	3.300	3.551	–0.251

Sources: Table 2.4a, R E Ebel, *Communist Trade in Oil and Gas*, New York, Praeger, 1970, p 138; *Vneshnyaya torgovlya SSSR, 1968*, pp 26, 213; *1969*, pp 26, 211; *1970*, pp 26, 215

Towards the end of the sixties increasing logistic difficulties obliged the Soviet Union to approach Iraq for oil supplies. Such imports could either be delivered via the Black Sea into the southern area of the Soviet Union to compensate for declining production in the Baku area, or directly to established export markets. The relatively late development of the Soviet gas industry coincided with the deterioration of economic, technical and logistic conditions in the oil industry. Though not without problems of its own the gas industry's potential enabled planners to make a substantial adjustment in forward estimates of energy supply and consumption.

In January 1969 the Soviet Oil Minister, V D Shashin, declared that because of rising domestic and Eastern European consumption, exports of Soviet oil to non-communist markets would not rise 'significantly'.[52] In interpreting this statement much depends on the meaning of the word 'significantly'. Ebel took this as indicating that the growth of net Soviet exports would be very slight in the future.[53] However, neither Ebel nor the Soviet Oil Minister mentioned the potential impact of gas.

In the aftermath of the Six Day War of 1967 and the closure of the Suez Canal, the Soviet Union, faced with increasing difficulties in supplying its own oil to established markets in the Far East and Japan, negotiated the exchange of oil and refined products at the Black Sea ports, for delivery to international oil industry outlets. In return an equivalent quantity was delivered at Persian Gulf ports, for delivery to Soviet customers. This had been common practice in the international oil industry but represented a new departure for the Soviet oil export authority Soyuznefteeksport. The convenience of such arrangements became increasingly important to the Soviet Union as the distribution of its own oil grew more costly.

Concluding his study, written in 1968, Robert Campbell noted that Soviet reserves of oil and gas would not constitute a constraint on future export levels, but that the prime factor was cost.[54] He identified the crucial issue, namely that Soviet gas might prove to be a more advantageous substitute for coal in the domestic energy balance, thus freeing oil for export. But he confined his forward view to stating that 'success in expanding gas output . . .(was). . . bound to have an important influence on oil export policy'.[55] At the outset of the ninth Plan the prospects for the success of such a policy seemed promising.

References

1. The circumstances leading to the decline in crude oil production are outlined in: (i) R E Ebel, *The Petroleum Industry of the Soviet Union*, Washington DC, American Petroleum Institute, 1961, pp 62-65; (ii) R W Campbell, *The Economics of Soviet Oil and Gas*, Baltimore, Johns Hopkins Press, 1968, p 122; (iii) R E Ebel, *Communist Trade in Oil and Gas*, New York, Praeger, 1970, pp 28-29.

2. Ebel, *Communist Trade . . .*, (1970), p 27.

3. Campbell, *op cit*, p 126.

4. *ibid*, pp 127-128.

5. V M Baryshev, in *Azerbaidzhanskoe neftyanoe khozyaistvo*, 12/1938, p 8.

6. I M Gubkin, 'Zadachi neftyanoi geologii na Yuzhno-vostochnom Kavkaze v tret'ei pyatiletke', *Azerbaidzhanskoe neftyanoe khozyaistvo*, 12/1938, p 12.

7. D B Shimkin, *The Soviet Mineral-Fuels Industries 1928-1958: A Statistical Survey*, Washington DC, US Government Printing Office, 1962, p 41.

8. V A Kalamkarov, 'Rezervy dal'nego razvitiya neftyanoi promyshlennosti', *Neftyanoe khozvaistvo*, 5/1955, p 2.

9. Campbell, *op cit*, p 126.

10. V P Sukhanov, 'Puti razvitiya neftepererabatyvayushchei promyshlennosti v SSSR', *Neftyanoe khozyaistvo*, 9/1958, p 2.

11. Ebel, *The Petroleum Industry . . .*, (1961), p 65.

12. Shimkin, *op cit*, p 45.

13. A Nove, *An Economic History of the USSR*, Harmondsworth, Penguin, 1972, pp 315-316.

14. In fact the coal industry failed to break even until the price reform of 1967 (see *Narodnoe khozyaistvo SSSR 1968*, p 745; *1970*, p 705). On the question of prices for oil products, industrial consumers, for example, paid 393 rubles per tonne for lighting kerosene, general consumers 782: the Oil Ministry charged its own transport network 40 per cent less than other consumers for all types of fuel and lubricant. Shimkin, *op cit*, p 44.

15. *ibid*, p 45.

16. Campbell, *op cit*, p 18.

17. P Agukin, A Shakhmatov, 'Neftyanaya i gazovaya promyshlennost' v shestom pyatiletii', *Planovoe khozyaistvo*, 3/1956, p 32.

18. *ibid.*

19. *ibid*, p 40.

20. *ibid.*

21. *ibid*, p 41.

22. Nove, *op cit*, pp 352-353.

23. *ibid*, p 353; *Narodnoe khozyaistvo SSSR 1975*, p 204; Agukin, Shakhmatov, *op cit*, p 41.

24. Campbell, *op cit*, p 125.

25. M M Brenner, *Ekonomika neftyanoi i gazovoi promyshlennosti*, Moscow, Nedra, 1968, p 31.

26. *ibid*, p 36.

27. I F Elliot, *The Soviet Energy Balance*, New York, Praeger, 1974, pp 36-37.

28. A V Aleksandrov, 'Osnovnye tendentsii nauchno-tekhnicheskogo progressa v gazovoi promyshlennosti, *Gazovaya promyshlennost'*, 1/1970, pp 25-26; Elliot, *op cit*, p 37.

29. Nove, *op cit*, p 353; see also Jeremy Russell, *Energy as a Factor in Soviet Foreign Policy*, Farnborough, Saxon House, 1976, p 74.

30. An analysis of the performance of the Soviet oil and gas industries during the eighth Plan is given by J R Lee in 'The Fuels Industries' in *Economic Performance and the Military Burden in the Soviet Union*, Washington DC, US Congress, Joint Economic Committee, 1970, pp 33-37.

31. Agukin, Shakhmatov, *op cit*, p 35.

32. Reported in V S Varlamov, 'Problemy transportnogo osvoeniya Zapadno-Sibirskoi nizmennosti v svyazi s formirovaniem na ee territorii novogo narodnokhozyaistvennogo kompleksa', *Izvestiya AN SSSR, Ser. geograficheskaya*, 3/1967, p 48.

33. Lee, *op cit*, p 35.

34. *ibid.*

35. Elliot, *op cit*, p 47.

36. N V Mel'nikov, 'Voprosy razvitiya toplivnoi promyshlennosti', *Voprosy ekonomiki*, 1/1969, p 17.

37. Ebel, *Communist Trade . . .*, (1970), p 33.

38. P R Odell, *Oil and World Power: Background to the Oil Crisis*, Harmondsworth, Penguin, 1975 (4th ed.), p 52.

39. Discussed in some detail in Nove, *op cit*, pp 348-350.

40. The background to the Soviet Union's trade agreement with Cuba is given in Odell, *op cit*, pp 59-62.

41. See M Tanzer, *The Political Economy of International Oil and the Underdeveloped Countries*, London, Temple Smith, 1970, pp 178-193.

42. National Petroleum Council of America, *The Impact of Oil Exports from the Soviet Bloc* (2 vols), Washington DC, American Petroleum Institute, 1962. A supplement to the report was added in 1964.

43. *ibid*, vol 2, p 418.

44. For example, when the Suez Canal was closed in 1967, resulting in disruption in supply from the Middle East, and oil shortages in Western Europe and Japan, the Soviet Union raised its oil prices in these markets. Ebel, *Communist Trade . . .*, (1970), p 63.

45. *ibid*, p 64.

46. P J D Wiles, *Communist International Economics*, Oxford, Basil Blackwell, 1968, p 40.

47. The background to this agreement is given in Ebel, *Communist Trade . . .*, (1970) pp 135-137.

48. *ibid*, p 138.

49. *ibid*, pp 140-142, 149-155.

50. *ibid*, pp 155-163.

51. *ibid*, p 164.

52. Reported in *New York Times*, 11 January 1969, p C-39.

53. Ebel, *Communist Trade . . .*, (1970), p 104.

54. Campbell, *op cit*, p 251.

55. *ibid*, pp 252-253.

Chapter 3

Soviet Oil and Gas, 1971-1975

There is a marked difference in the development of the Soviet energy economy in the ninth (1971-1975) Five-Year Plan compared with the previous 15-year period. From 1955 to 1965 the growth in the share of oil and gas in the Soviet energy balance was brought about by the development of reserves located in the European part of the USSR, relatively close to centres of consumption, namely the Volga-Ural region, the North Caucasus and the Ukraine. From 1955 to 1965 production of oil and gas condensate in the Soviet Union grew by 192.1 million tonnes, of which European areas accounted for 150.3 million, 87 per cent of the total growth. In the same period gas production grew by 118.7 billion cubic metres, over 70 per cent of which was provided by European sectors, mainly the same areas which accounted for the bulk of growth in oil production.[1] The position in the coal industry differed from that of oil and gas in that between 1955 and 1965 less than half (44 per cent) of additional production was derived from European operations. The overall position was that European Russia's contribution to total fuel production rose from 66 per cent in 1955 to 72 per cent in 1965.[2]

Development prospects and plans for the 1971-1975 period showed the increasing importance for the Soviet energy economy of regions to the east of the Urals. Opportunities for increasing the production of oil and gas in the European part of the Soviet Union were seen to be limited by a declining rate of discovery of new reserves. The objective was therefore to maintain or slightly increase production levels in fields operational during the eighth Plan, though in certain areas production was scheduled to decline. In European Russia the areas that were scheduled to increase oil production were certain fields in the Urals, the North-West and new fields in Belorussia.[3] The European area as a whole was scheduled to provide one fifth of the increased production during the ninth Plan, with the balancing four fifths to be provided by operations in West Siberia,[4] where the production level in 1975 was planned to be 120 to 125 million tonnes.[5] Table 3.1 details the planned changes in regional production of oil (excluding gas condensate) in 1975, compared with 1970.

The development of the oil industry in the ninth Plan was paralleled by that of the gas industry. In the course of the ninth Plan, one fifth of the increase in production was scheduled to be provided by European operations, the balance being provided by West Siberia and Central Asia. The rapid development of gasfields in the Komi Autonomous Republic and the Orenburg oblast', discovered and prepared for development during the eighth Plan, was expected to be

Table 3.1 Changes in the Regional Pattern of Soviet Crude Oil Production, 1975 Plan compared with 1970 actual (million tonnes)

	1970 Production	*1970* % share	*1975P* Production	*1975P* % share	*1975P as % of 1970*
USSR	348.8	100.0	496.0	100.0	142.2
European Area plus Urals	285.2	81.8	314.5	63.4	110.3
of which:					
Tatar ASSR	101.9	29.2	101.0	20.4	99.1
Bashkir ASSR	39.2	11.2	40.0	8.1	102.0
Kuybyshev oblast'	35.0	10.0	35.0	7.1	100.0
Orenburg oblast'	7.4	2.1	14.0	2.8	189.2
Perm oblast'	16.1	4.6	21.5	4.3	133.5
Komi ASSR	5.6	1.6	10.0	2.0	178.6
Regions to East of Urals	63.6	18.2	181.5	36.6	290.0
of which:					
West Siberia	31.4	9.0	125.0	25.2	400.0
Turkmen ASSR	14.4	4.1	22.0	4.4	152.8
Kazakhstan	13.1	3.8	30.0	6.0	230.0

Source: *Pyatiletnii plan 1971-1975*, p 103

Table 3.2 Changes in the Planned Regional Production of Gas, 1975 compared with 1970 (billion cubic metres)

	1970 Production	*1970* % share	*1975P* Production	*1975P* % share	*1975P as % of 1970*
USSR	198.0	100.0	320.0	100.0	161.6
European Area plus Urals	139.0	70.2	164.1	51.3	118.0
of which:					
Orenburg oblast'	1.3	0.7	26.0	8.1	X 20.0
Komi ASSR	6.9	3.5	16.1	5.0	X 2.3
Regions East of Urals	59.0	29.8	155.9	48.7	X 2.6
of which:					
West Siberia	9.3	4.7	44.0	13.8	X 4.7
Turkmen ASSR	13.1	6.6	65.1	20.3	X 5.0

Source: *ibid*, pp 106-107

Table 3.3 Planned Soviet Oil and Gas Production 1971-1975

	1971	*1972*	*1973*	*1974*	*1975*
Oil, excluding gas condensate (million tonnes)	371.3	395.1	429.0	461.0	496.0
Natural Gas (billion cubic metres)	211.0	229.0	250.0	280.0	320.0

Source: *ibid*, p 346

counterbalanced by a decline in production in other European fields.[6] Table 3.2 details the planned changes in regional production of gas in 1975 compared with 1970. The year-by-year growth pattern planned for the oil and gas industries for the term of the ninth Plan is outlined in table 3.3.

In the oil refining industry the objective was to increase the capacity of individual processing units rather than to instal large-scale new refineries. It was stated in the Plan that the average capacity of primary distillation units should rise by 62.5 per cent from 3.2 million tonnes per year recorded in the eighth Plan to 5.2 million in 1971-1975; that catalytic power-formers, used to manufacture motor gasoline and petrochemical feedstock, should increase from an average capacity of 384 thousand tonnes per year during the eighth Plan to 566 thousand; and that hydrofiners, which are used to remove sulphur from middle distillate fuels, should increase from 1.12 million tonnes per year to 1.85 million.[7] The total oil refining capacity was scheduled to rise by 40 per cent during the ninth Plan,[8] which required an increase in capital investment of 62.1 per cent.[9] In order to achieve these objectives, refinery capacity operational at the end of the eighth Plan would need to be increased by 20 per cent, the balance being provided by the construction of new refineries at Lisichansk (Ukraine), Mozyr (Belorussia), Jurkarkas (Lithuania), Arkhangelsk (on the White Sea), Pavlodar and Chimkent (Kazakhstan), Chardzhou (Turkmenia), Achinsk (West Siberia) and Khabarovsk (Far East). Given that some 70 per cent of new capacity was scheduled to be provided by the four refineries planned for European Russia and around 30 per cent by the remaining five Asiatic plants, it is clear that the capacity of the European refineries would be some three times larger than the Asian.[10]

As the centre of exploration and production moves eastwards, long-range criteria in integrating oil and gas into the total energy balance differ from short- and medium-range criteria. Therefore decisions on minimum financial outlay taken at the initial stage of development do not serve as the sole basis for the evaluation of available alternatives. Factors taken into consideration by Soviet planners include estimates of forward price parities, reliability of supply, feasibility of demands made by the fuel and energy sectors on related industries such as machine-manufacturing, civil engineering and metallurgy, and the impact of energy development on the environment.[11]

However, the basic factors that influence the formulation of energy policy are the discovery of new fuel reserves, judged economic on the basis of the best available information, plus the cost of transporting fuels in processed form to the point of final consumption. By the mid-point of the ninth Plan the targets for the discovery of new oil reserves had been underfulfilled by some 50 per cent.[12] This necessitated some degree of rethinking on the part of Soviet planners as to the general future direction of the energy economy. It is important to note that the short-term impact of delays in the exploitation of reserves declared to be economically attractive goes beyond that of simple shortages of a given fuel. Thus, for example, delays in bringing on stream oil reserves scheduled for development during the ninth Plan, combined with a refining policy aimed at maximizing the output of non-substitutable fuels and petrochemical feedstock, caused a shortage of residual fuel oil.[13] This stimulated a demand by industrial consumers for change in the priorities of the refinery balance. The longer-term impact of a delay of this nature is that of obliging planners to direct a greater proportion of natural gas to fuel end-uses than had been previously considered optimal. In so doing the supply of gas as petrochemical feedstock is tightened. Towards the end of the ninth Plan the problems encountered in the oil industry were compounded by the commencement of similar shortfalls in the discovery of new reserves of other fuels.[14]

A recent feature of Soviet development policy has been to limit the construction and expansion of energy-intensive industry in European Russia and to site such enterprises closer to the new centres of fuel production. According to Gosplan's Committee for the Study of Productive Resources, fuel can then be delivered to the enterprise at between one fifth and one half the cost of delivery in European Russia.[15] This is gradually becoming the preferred policy of Gosplan. It acknowledges that in the short term it would be difficult to eliminate the fuel and energy deficit in European Russia by expanding fuel production there, given the long lead-times in construction[16] and a wide-spread disparity in the delivered cost of fuels (see table 3.4). Two prominent Soviet energy specialists defined the problem of the Soviet energy economy as an interaction of three factors: the concentration of productive capacity, creating pockets of very high energy demand; the definition and achievement of an optimal variant amongst available fuels; the intrinsic inertia and changing capital intensity of the fuels industries.[17] Prior to analyzing in detail the development of the Soviet oil and gas industries during the ninth Plan, it is useful to consider the current system of energy planning as a whole, in order to establish the background to decisions taken in respect of the development of these two industries.

Table 3.4 Comparison of Cost of Fuels Delivered to Selected Consumption Points in European Russia (rubles per tonne of standard fuel)

Consumption Point	*Donets Coal*	*Kuznetsk Open-cast Coal*	*Tyumen' Gas*	*Central Asian Gas*	*Fuel Oil*	*Peat*
Leningrad	20.3	17.3	10.8	–	4.4	–
Moscow	18.4	16.1	10.0	13.3	8.1	11.6
Minsk	18.6	–	11.4	–	9.3	10.2
Gor'kii	18.8	15.0	10.0	–	7.7	–
Saratov	17.5	15.2	13.3	10.4	7.3	–
Donetsk	15.6	–	–	12.9	10.8	–

Source: A E Probst, 'Puti razvitiya toplivnogo khozyaistva SSSR', *Planovoe khozyaistvo*, 6/1971, p 56

Table 3.5 details the planned fuel production targets for 1971 to 1975 and the results achieved. Targets for an increase in the discovery of new fuel reserves were set at 33.7 per cent, but no precise figure was given.[18] The Plan envisaged that the share of oil and gas in the overall balance would rise from 60.4 per cent in 1970 to 67.4 per cent in 1975.[19] Taking into account the directive to increase the share of open-cast mined coal from 26.7 to 30.9 per cent of total coal produced, Soviet planners expected the use of 'economic fuels' ('ekonomicheskie vidy topliva') to rise from 68 to 75 per cent of fuel consumed in this period.[20]

Table 3.5 Soviet Fuel Production 1971-1975, Plan and Performance

		1971	*1972*	*1973*	*1974*	*1975*
Oil (million tonnes)	Plan	371	395	429	461	496
	Actual	372	394	421	451	491*
Gas Condensate (million tonnes)	Plan	na	na	na	na	9
	Actual	5.3	6.6	7.6	8.3	na
Gas (billion cubic metres)	Plan	211	229	250	280	320
	Actual	212	221	236	261	289
Coal (million tonnes)	Plan	620	634	652	670	695
	Actual	641	655	668	685	701

na = not available

*Note that *Narodnoe khozyaistvo SSSR 1975* did not record separate figures for production of oil and gas condensate: this figure includes both

Sources: *Pyatiletnii plan 1971-1975*, pp 28, 98, 346; *Pravda* 1 February 1976, p 1; *Narodnoe khozyaistvo SSSR 1974*, p 183; *Narodnoe khozyaistvo SSSR 1975*, p 205

Despite a preference for increasing the share of oil and gas within the energy balance, coal has remained a significant contributor, notably in electricity generation. This is particularly the case in European Russia.[21] In the Soviet Union as a whole electricity generation accounts for approximately one third of energy consumption, and some 60 per cent of electricity is generated in coal-fired power stations.[22]

Since the ratio of proved reserves to production is less favourable for oil than for other conventional fuels and the transportation problems more severe, oil has come to be regarded as the scarce resource in relation to which the energy balance as a whole should be optimized. On the one hand a certain quantity of light petroleum fractions must be produced for which no ready substitute exists, either in the technical or economic sense.[23] On the other hand there is a substantial degree of choice amongst boiler and furnace fuels and the option taken depends on the economic balance between production, transportation and conversion costs.[24] The decision-making process by which the final fuel balance is determined lies within the remit of the Collegium of Gosplan.[25] The final requirement is worked out by aggregating reported demand by sector of the economy. Individual departments of Gosplan are responsible for verifying the sectoral analysis of energy demand: the Gosplan Collegium accordingly aggregates sectoral demand and allocates fuel.[26] Mathematical models are used to plan at three levels, the long-term (the next Five-Year Plan), the medium (the remainder of the current Plan) and the annual Plan.[27] It appears that the use of linear programming techniques is a relatively recent feature of Soviet energy planning, and that work on the development of an all-Union unified energy supply model is still incomplete.[28]

One of the early detailed works on the Soviet method of energy planning was that of the German economist Werner Gumpel.[29] In his analysis he pointed to four major difficulties faced by Soviet planners: first, the worsening spatial dislocation between energy reserves, production and consumption; second, the changing criteria determining levels of substitutability between fuels; third, the fluctuations in consumption patterns caused principally by the continental climate; and fourth, the long lead-times for investment projects in the energy industries, which are often subject to overruns.[30] Gumpel saw the last of these features as the crucial issue. Perhaps the most important issue is that of the uncertainty and changeability of the basic information upon which decisions are made.[31] During the ninth Plan it was felt that forward estimates on reserves and production costs of coal were more reliable than those for oil and gas. More precisely, Gosplan's information was apparently less firm in respect of the comparative costs of production and transportation of Siberian oil and gas resources than of the costs relating to the future exploitation of Kansk-Achinsk and Kuzbass coal.[32] Thus, as Melent'ev and Makarov argued,[33] it became optimal to favour an energy balance which slowed the rate of increase of oil and gas in relation to coal, on the grounds of minimum uncertainty. On the other hand planners have to take into account the economics of conversion and the foreign trade implications of domestic production decisions. This became more urgent after 1973, given the enhanced earning potential in hard currency markets afforded by the increased world price for oil. In consequence, there has been a greater interest on the part of Soviet planners in coal and gas as fuels for electricity generation,[34] and in gas partly as fuel and partly as feedstock for a range of industries.[35] Therefore gas and coal can be used as substitute fuels for oil. The relative decline in the share of the refining balance held by fuel oil permits a correspondingly increased share of naphtha and middle distillate output, for which no ready substitute exists and which are in greater demand than fuel oil in export markets.

The optimum utilization of fuel is a further objective of Soviet energy policy. On the basis of its conversion characteristics natural gas is generally the most efficient of the three major primary energy sources.[36] In the course of the ninth Plan there was a pronounced emphasis on natural gas as the preferred fuel in the long term for both domestic and industrial use.[37] The 1975 consumption balance for gas is outlined in table 3.6. The striking feature of this balance is the low percentage share of gas directed to petrochemical manufacture and to export. These are the two sectors that Soviet planners most wish to develop in the immediate future. Conversion efficiency in general has become the subject of some attention in Soviet writing: it is admitted that despite considerable improvement over the 25 years to 1975 energy conversion efficiency is unacceptably low.[38]

In concluding his study Elliot observed that the Soviet Union 'has no need to fear an absolute energy shortage'.[39] Indeed there is no evidence to refute this statement, and the same could be said of a number of the world's industrialized countries. The point at issue is the nature, extent and implications of the politico-economic problems that are known and admitted to exist. Soviet analysts have been at pains to point out that the Soviet Union does not have an energy supply problem, nor is the country affected by the type of 'energy crisis' that has adversely affected the capitalist world.[40] If the Soviet Union, and the Comecon bloc as a whole, can remain economically self-sufficient in energy,

Table 3.6 Natural Gas Consumption by Economic Sector 1975

Sector	*% share*
Electricity generation	14-15
Heat raising	33
Industrial manufacturing	29
Petrochemical feedstock	7-8
Domestic fuel	5-6
Agriculture	0.5
Other	1
Export	balance

Source: E Yudin, 'Effektivnoe ispol'zovanie toplivno-energeticheskikh resursov strany – obshchego-sudarstvennaya zadacha', *Planovoe khozyaistvo*, 6/1975, p 61

particularly in oil, the area would be less vulnerable to the economic and political consequences arising from the oil price rises imposed by OPEC, and to allied rises in the price of other energy forms that might have to be imported.

As late as June 1973 one Western analyst, Professor Robert Campbell, was discussing as a major issue in Soviet energy policy, whether the USSR could take advantage of 'cheap Middle East oil', or 'allow' Eastern Europe to do so.[41] The very rapidity with which the trading conditions of the world market changed, and the extent to which the new pricing levels affected the Soviet Union and Eastern European view of import possibilities were such, that by 1976 the question of how Comecon countries would be able to finance the import of OPEC oil at levels already reached was raised by Western analysts.[42]

The Soviet Oil Industry 1971-1975: Objectives and Achievements

The oil industry's objective was to raise production of oil and gas condensate from 353 million tonnes in 1970 to 505 million in 1975. This was to be done by maintaining production levels achieved during the eighth Plan in the Volga-Ural oilfields, and by developing new deposits in West Siberia, the Komi Autonomous Republic, the Northern Caucasus, the Perm and Orenburg oblasts. the Udmurt Autonomous Republic, the Ukraine, Turkmenistan and the Caspian shelf.[43] Of these newly developed deposits West Siberia was planned to account for a growth of 90 to 100 million tonnes, Kazakhstan approximately 17 million, Turkmenia, the Orenburg oblast', the Komi ASSR and Belorussia approximately 30 million.[44] The eastward shift in explored ('razvedannye') reserves that influenced the setting of these priorities is detailed in table 3.7, in which gas reserves are also included.

The question of the extent of Soviet oil reserves and of success in their discovery is a prime factor in determining production policy overall, and also the regional distribution of production. Most Western analysts have pointed to the fact that since 1947 information on Soviet oil reserves has been regarded as a state secret.[45] However, a number of estimates of size and potential do exist.

Table 3.7 Geographical Shift in Location of Explored Reserves (%)

Area	1960		1973	
	Oil	*Gas*	*Oil*	*Gas*
USSR	100	100	100	100
European Russia	96	68.1	51	19.5
Eastern Regions	4	31.9	49	80.5

Source: Yu I Bokserman, 'Nekotorye tendentsii dal'neishego razvitiya toplivnoi promyshlennosti', *Neftyanik*, 1/1975, p 7

One estimate puts explored reserves at 4.5 billion tonnes,[46] another at 3 billion,[47] the latter figure giving a ratio of explored reserves to production of 10 to 1 at the time of the estimate. One Soviet source noted that whilst production doubled during the years 1960 to 1968, discovery of explored reserves increased by only 51 per cent.[48] Elliot expressed the view that since over 80 per cent of reserves lay at depths of less than 2,000 metres, extraction conditions were very favourable.[49] Even in the absence of precise data, Western analysts agree that the level of exploitable reserves currently known to exist, will not constitute a constraint on the levels of production envisaged in the medium-term.[50]

However, as a result of developments in the oil exploration sector from 1960 to 1975 the location of reserves has undergone a substantial shift. During this time total oil reserves have doubled in European Russia, increased twelvefold in the Volga-Ural area and fifteenfold in regions to the east of the Urals.[51] Granted that in the short term the major issue is not the extent of total reserves, but of economic reserves with supporting logistic systems, there were still in 1975 significant reserves in European Russia, with the prospect of total reserves substantially above likely production levels. These were concentrated in small fields, in contrast to Siberian reserves, which are for the most part to be found in extremely large fields.[52]

The importance of the development of West Siberia for the Soviet economy generally was stressed at the outset of the ninth Five-Year Plan. It was thought possible that West Siberia could be developed at a rate that would result in Siberia's providing as much as 30 per cent of the Soviet Union's energy requirement in 1980, rising to 40-45 per cent by the year 2000.[53] Table 3.8 details the changing importance of established oil-producing areas for Soviet oil production as a whole, illustrating the peak and decline of production in the Volga-Ural fields and the emergence of West Siberia (grouped here with offshore production around the Mangyshlak peninsula in the Caspian Sea). In addition to the fact that oilfields discovered to date in West Siberia have been particularly large, production costs are the lowest in the Soviet Union, as detailed in table 3.9.[54]

However, these advantages have been offset to some extent by the difficulties encountered and expense incurred in meeting drilling targets for new producing wells. This was especially the case in the Tomsk and Tyumen' oblasts, which along with the Middle Ob' area constitute the major proportion of West Siberian production and potential.[55] It has also been pointed out that, in theory, the siting of refineries and petrochemical plants in West Siberia and the onward transportation of finished products can show capital cost savings of 12 to 14 per

Table 3.8 Crude Oil Production in Geographic Zones, 1970-1975 Plan (million tonnes)

Zone	*1970 Production*	*%*	*1973 Plan Production*	*%*	*1975 Plan Production*	*%*
European Russia plus Urals	285.2	81.8	300.3	70.0	314.5	63.4
of which: Volga	184.4	52.9	188.1	43.8	182.6	36.6
Eastern Regions	63.6	18.2	128.7	30.0	181.5	36.6
of which: West Siberia plus Mangyshlak	41.8	12.0	104.9	24.5	155.5	31.3
Total USSR	348.8	100	429.0	100	496.0	100

Sources: *Table 3.1*; A E Probst, Ya A Mazover (eds), *Razvitie i razmeshchenie toplivnoi promyshlennosti*, Moscow, Nedra, 1975, p 88

Table 3.9 Comparison of Oil Production Costs (All-Union average = 1)

Zone	*Relative Cost*
European Russia plus Urals	1.1 – 1.2
Siberia and Far East	0.7 – 0.8
Central Asia and Kazakhstan	1.5

Source: *ibid*, p 91

Note that the characteristics of the three energy zones are stated to be as follows:
(i) European regions plus the Volga-Ural area have high consumption and the main component of the regional fuel balance is Donetsk, Moscow and Pechora coal, the highest cost fuel in the Soviet Union. Despite the availability of cheap hydroelectricity, this zone has the highest electricity cost in the country.
(ii) Siberia and the Far East contain, at present, the majority of Soviet energy reserves and have very low energy consumption.
(iii) Central Asia and Kazakhstan are self-sufficient in energy and abundant in gas, and are substantial contributors to the energy supply of zone 1.
Characteristics of these 'energy zones' are discussed in detail in K M Zvyagintseva, 'O trekh zonakh toplivno-energeticheskogo khozyaistva Sibiri', *Izvestiya SO AN SSSR Ser. obshch. nauk*, 1/1974, pp 19-25

cent, compared with the siting of the similar complexes in European Russia and transportation there of energy and feedstock.[56]

A further issue affecting achievements and prospects in West Siberia is the fact that the main administrative organization, Glavtyumenneftegaz, has to coordinate the efforts of production enterprises sited at a great distance from one another. The average distance from the administrative center in Tyumen' is 300 to 400 kilometres and some fields are as much as 1,000 kilometres apart.[57]

In an analysis written at the end of 1974 it was estimated that the initial oil production target for 1975 of 125 million tonnes for West Siberia would be exceeded.[58] The final figure was in fact 148 million tonnes.[59] Production of oil in West Siberia in 1970 totalled 31.4 million tonnes.[60] The success experienced in raising production during the ninth Plan led planners to set a target for the area of 300 to 310 million tonnes in 1980.[61]

The size of fields discovered to date in West Siberia and their relatively low production costs are not the only feature that has prompted their rapid development. The chemical characteristics of the available oils are particularly well suited to the changing priorities of both the Soviet fuel and petrochemical industries, in that the majority of West Siberian crude oils are of light gravity, having a high yield of gasoline, naphtha and middle distillates.[62] Moreover there is no relative loss of lubricant basestock and bitumen: the relative loss is in the percentage yield of fuel oil, which is the most readily substitutable product.[63]

However, the development of the oil reserves of West Siberia has not been devoid of problems. In the early part of 1973 an article appeared suggesting that Gosplan was unhappy about the return on investment in Siberia in general and in Tyumen' in particular.[64] The author of the article, V Bogachev, emphasized that Siberian construction costs were running at an average of 30 to 40 per cent, and in Tyumen' at 50 to 100 per cent, above original estimates.[65] In Bogachev's view the main reason for the overrun on estimated cost was that all areas to the south of the 60th parallel, including West Siberia, were grouped for the purposes of cost estimation and budgeting in the same 'cost belt' ('poyas') as the Baltic area and European Russia. However, in reality, wages and material transportation costs are substantially above the all-Union average and this resulted in under-budgeting for many of the sites, with consequent delays in completion of projects and cost overruns.[66] Bogachev argued that it was unrealistic not only to include West Siberia in the aforementioned cost belt, but also to attempt rigid advance programming of levels of investment in individual sectors of the Siberian development Plan.

The uncertainty of information is the most pressing issue in exploratory operations in this area.[67] This has been complicated by the fact that initial achievements in the Tyumen' oblast' were particularly impressive and augured well for future activity. For example, the rate of discovery of oil reserves per metre of drilling in Tyumen' in the late sixties was six times the all-Union average.[68] However, in the course of the ninth Plan the technical difficulties experienced in the development of West Siberia were more severe than had been anticipated, with the result that drilling performance deteriorated.

In the Spring of 1969 a conference was held in Tyumen' to discuss short-term industrial development, with special reference to the oil industry. Central to the discussion was the question of determining an optimal oil production level for West Siberia to 1980. Estimates of this optimal level ranged from that of the oil industry ministry and management of Glavtyumenneftegaz, who advanced a figure of 75 to 80 million tonnes, to that of the Deputy Minister of Geology of the RSFSR, who put forward a figure of 150 million tonnes.[69] The consensus of the conference was unclear. On the one hand the section of the conference report entitled 'Geological Exploration' revealed an estimated production level in West Siberia of 150 to 200 million tonnes by 1975, whereas the section entitled 'Oil Industry' gave an estimated production level of 70 to 100 million tonnes in 1975, rising to 180 to 200 million in 1980.[70]

The differences between these early estimates, the final performance in the ninth Plan and the target set for the tenth, serve to put the development of the West Siberian oil industry during the ninth Plan into perspective. The inability to maintain production levels in European Russia and the Volga-Ural fields during the ninth Plan necessitated an acceleration of the rate of development of West Siberia to a degree that Siberian operations accounted for almost the entire net increase in production.[71] The relative importance of West Siberia to total Soviet production, and in particular the contribution of the Samotlor oilfield, is outlined in table 3.10. In fact, the discovery and development of the Samotlor oilfield has been one of the principal features of the oil industry as detailed in the ninth Plan, and has been the subject of recent articles by F G Arzhanov and Yu B Fain,[72] and by R I Kuzovatkin.[73]

Table 3.10 Siberian Contribution to Soviet Oil Production 1971-1975 (million tonnes)

Production	*1971*	*1972*	*1973*	*1974*	*1975*
Total USSR	377	400	429	459	491
West Siberia – Plan	44	61	82	104	125
– Actual	45	63	88	118	148
Tyumen' oblast'	28	na	na	111.4	143.2
Samotlor Oilfield	10	21.2	39	61.2	86.5

(Figures include gas condensate) (na = not available)

Sources: *Narodnoe khozyaistvo SSSR 1974*, p 183; *Pravda*, 1 February 1976, pp 1, 2; *Ekonomika i organizatsiya promyshlennogo proizvodstva*, 5/1976, p 7, and 6/1976, p 80; *Neftyanoe khozyaistvo*, 5/1975, p 39; *Ekonomicheskaya gazeta*, 19/1975, p 6

Despite the severe climate and the difficulty in communications there has been a rapid development of the Samotlor field. This has been achieved through the application of the technique of cluster drilling, a process by which several wells can be drilled simultaneously from a single point, thus facilitating high-speed extraction of the crude oil from the deposit.[74] Geologists estimate that the Samotlor field is the Soviet Union's largest and that it is likely to have a life of at least 10 years. In 1975 a production peak of 120 million tonnes per year was predicted for this field.[75]

However, this peak will be reached relatively quickly, and output from the Middle Ob' area, including that from Samotlor, is expected to decline in the eighties.[76] Consequently during the ninth Plan exploratory effort began to shift to the more remote oil fields of North Tyumen', development of which was expected to be fraught with severe logistic problems and rising costs.[77]

Cost escalation in the oil industries of Central Asia and Kazakhstan has been a problem during the ninth Plan. The area is predominantly one of mature fields, all of which have passed production peak and which are in the process of decline. In Turkmenia, the rate of oil extraction had begun to exceed the rate of discovery of new reserves as early as 1970,[78] but production was nonetheless scheduled to rise from 15.4 million tonnes in 1970 to 22 million in 1975.[79] However, the oilfields around the Mangyshlak peninsula became the major area of importance in Central Asia and Kazakhstan during the ninth Plan. Mangyshlak operations were scheduled to provide 90 per cent of the increase in production sought from Kazakhstan from 1971 to 1975.[80] By the middle of this period it became evident that the production association Mangyshlakneft' was facing difficulties. The 1972 production figure of 15 million tonnes was below target due to failure on the part of the construction sector to keep pace with the installation requirements. Mangyshlakneft' itself was the source of some of the problems: disparate rates of production between wells, due to deposits in the area tending to be dispersed over a number of geological levels, compounded with substantial seasonal fluctuations in production, accounted for the shortfall.[81]

At the outset of the ninth Plan it was intended to raise production in the European zones of the Soviet Union, predominantly in the Ural area (the Udmurt ASSR and the Perm and Orenburg oblasts), the Komi ASSR and Belorussia, in accordance with the data presented in table 3.11. Production in the Udmurt ASSR commenced in 1969,[82] and by 1974 an output of 2.75 million tonnes was anticipated.[83] However, the Udmurt fields are particularly difficult to exploit:

Table 3.11 Anticipated Change in Oil Production in European Russia, 1971-1975 Plan

	1970		*1975 Plan*	
	million tonnes	*% of USSR*	*million tonnes*	*% of USSR*
European Russia	265.2	76.2	295.6	59.7
of which:				
Volga-Ural	208.3	59.8	226.6	45.8
North Caucasus	34.0	9.8	34.1	6.9
Komi ASSR	5.6	1.6	10.0	2.0
Belorussia	4.2	1.2	8.5	1.7
Ukraine	13.1	3.8	16.4	3.3

Source: L I Suchkova, N M Faustova, V F Cherevadskaya, 'Effektivnost' razvitiya neftedobyvayushchei promyshlennosti evropeiskikh raionov strany v tekushchem pyatiletii', *Ekonomika neftyanoi promyshlennosti*, 10/1973, p 7

they are complex multi-layered structures, yield per well is low, the oil viscous and high in paraffin wax making transportation in pipelines more difficult. In addition the major fields lie in marshy terrain at some distance one from another.[84] The Perm oblast', one of the oldest producing areas in the Soviet Union, was scheduled to increase production from some 16 million tonnes in 1970 to 27 million in 1975[85] but production level in 1975 turned out to be 22.3 million tonnes.[86] The major problem facing production teams in the Orenburg oblast' has been that of maintaining seam pressure in relatively deep layers, a necessary measure in the extraction of approximately 80 per cent of the oil.[87] The importance of the Komi ASSR rests on the development of the Timano-Pechora hydrocarbon area. A particular feature of the oil of this area is its variety: both light and heavy types of crude oil can be extracted.[88] The bulk of this oil is, however, paraffinic-based with a high yield of light products at distillation and low in sulphur content.[89] Like the West Siberian crude oils, these oils are particularly suitable for the refinery balance that the Soviet Union began to favour in the course of the ninth Plan. In the period from 1971 to 1975 the Komi ASSR produced a total of 53 million tonnes of oil, and it has been estimated that on the basis of this level of production, a peak output of 25 million tonnes per year could be achieved by 1980.[90] There was a threefold increase in the discovery of new reserves in the Komi ASSR during the ninth Plan contrasting sharply with the position in the Soviet Union as a whole.[91] In Belorussia prospecting for oil commenced in 1961 and the first oil was struck in August 1964 with the discovery of the Rechitsa field.[92] Production has grown from some 40 thousand tonnes in 1965 to just over 8 million in 1975.[93] This is considered about the limit and a significantly higher volume of production is not expected.

Interest grew in the potential of offshore oil reserves in the Baltic, Caspian and Sakhalin areas during the period of the ninth Plan. Exploration in the Baltic had commenced in 1959, though by 1975 only 10 fields had been discovered. These deposits are relatively shallow (1,500 to 2,500 metres), and the oil is almost free of sulphur but the light fractions obtained in refining are of a disappointingly low octane.[94] Exploratory work carried out during the ninth Plan indicated that oilfields known to exist in the South-East of the Baltic, off the Latvian and Lithuanian coast might prove attractive.[95] It is, however, the oilfields in the Caspian Sea that are judged to be potentially the major contributors to offshore developments on the basis of work carried out in the early seventies.[96] Despite the fact that the majority of the Caspian deposits are at depths in excess of 3,500 metres, the fields are believed to be large enough to offset the high cost of drilling in potentially low unit production costs.[97]

Exploration commenced in the Sakhalin area as early as 1927, and to date has remained the sole oil-producing area of the Far East Economic Region. Sakhalin oil is of a quality that commands a premium in export markets: it has a high yield of light products and its fuel oil fraction is low in sulphur. Middle distillates and naphtha derived from Sakhalin oils are exceptionally well suited as feedstock for the emerging petrochemical industry of the Far East Region.[98] Despite the technical attractiveness of this crude oil Soviet planners are not providing for a marked increase in production from Sakhalin. Exploration and production costs are more than twice the all-Union average,[99] but, used locally as feedstock, Sakhalin oils are more economic than those transported from the European and Siberian fields.[100] However, larger-scale exploration and production, under joint schemes with American and Japanese participation, could alter Sakhalin's role in the medium- to long-term: this is discussed in Appendix A.

East Siberia is now considered an oil-bearing area of some significance, but during the ninth Plan activity was confined to geological surveys and exploratory drilling.[101] The oil that has been discovered resembles Sakhalin oil in respect of sulphur content and chemical composition. East Siberia's long-term potential was a subject of discussion as early as 1960,[102] and by 1968 it was thought that the area could contain almost half the total fuel and energy reserves of the Soviet Union.[103] However, towards the end of the ninth Plan these estimates had been revised downwards and a more modest share of total Siberian, and hence of all-

Union reserves was allotted to East Siberia.[104] Data on the distribution of the major primary energy reserves of West and East Siberia are given in table 3.12. As in the case of the development of Sakhalin, Soviet planners do not anticipate a marked impact on total production from East Siberian operations before the mid-eighties: however, Japanese-Soviet joint projects currently under discussion could alter the status of the area in the medium-term. These issues are likewise discussed in Appendix A.

Table 3.12 Distribution of Usable Fuel and Energy Reserves in Siberia (%)

	West Siberia	*East Siberia*
Hydroelectricity	15.0	85.0
Coal	69.3	30.7
Natural Gas	76.9	23.1
Total Fuel	76.6	23.4

Source: K M Zvyagintseva, 'O trekh zonakh toplivno-energeticheskogo khozyaistva Sibiri', *Izvestiya SO AN SSSR Ser. obshch. nauk*, 1/1974, p 21

As calculated from table 3.5 the cumulative underfulfilment of the ninth Plan in respect of production of oil and gas condensate was some 30 million tonnes. The first year of the Plan in which an underfulfilment was recorded was 1972, and this resulted in a revision of the original 1973 target to 424 million tonnes, a figure which in its turn was underfulfilled. The original 1974 target of 461 million tonnes was revised to 452 million and was fulfilled by a narrow margin: the 1975 target was revised from 505 million tonnes to 489, and this was exceeded by 2 million. Table 3.13 summarizes the revisions to the original plan targets, and table 3.14 records the trend in availability of crude oil and gas condensate in the Soviet Union from 1971 to 1975.

In addition to the difficulties encountered in oil exploration and production numerous problems in associated industries have contributed to the shortfall in production against the original Plan. There have been periodic bottlenecks in the supply of equipment both for exploration and production. Towards the end of the Plan the Soviet Oil Minister stressed that despite a high level of success in developing improved equipment for the oil industry, manufacturers had failed repeatedly to produce and deliver an adequate quantity. Although new designs for drilling bits had been perfected, few had reached the production stage. The quality of drill-pipe was criticized and the under-supply of equipment for low-temperature exploration was cited as a major reason for the difficulties encountered in meeting targets for exploratory drilling and reserve discovery in West Siberia.[105] The problem of success indicators in management control and assessment had itself caused some of the problems: for example, the criterion for Plan fulfilment was the number of metres drilled, with the result that on occasions drillers experiencing technical difficulties would leave a well unfinished and commence relatively easy drilling elsewhere in order to meet their target.[106] Difficulties had also been encountered in providing pipeline in the required quantities to support development, especially in West Siberia,[107] and there had

Table 3.13 Original and Revised Annual Plans and Performance of the Soviet Oil Industry 1971-1975 (million tonnes)

	1971	*1972*	*1973*	*1974*	*1975*
Original Plan: Oil	371	395	429	461	496
Condensate	na	na	na	na	9
Revised Plan (Oil and Condensate)	–	–	–	451	489
Actual Production (Oil and Condensate)	377	401	429	459	491

na = not available

Sources: *Table 3.5*; A Nove in *ABSEES, July 1974*, p xviii

Table 3.14 Availability of Oil in the Soviet Union 1971-1975 (million tonnes, oil and oil products, including gas condensate)

	1971	*1972*	*1973*	*1974*	*1975*
Domestic Production	377	400	429	459	491
Plus Imports	7	9	15	5	8
Less Exports	105	107	118	116	130
Availability	279	302	326	348	369

Sources: *Narodnoe khozyaistvo SSSR 1975*, p 205; *Vneshnyaya torgovlya SSSR 1972*, pp 27, 41; *1974*, pp 29, 44; *1975*, pp 25, 40

been a high labour turnover in both Mangyshlak,[108] and West Siberia[109] despite the significant privileges granted to oil and gas industry workers in the latter area.[110]

On 31 March 1975 a conference was convened in Moscow to discuss methods of raising efficiency in oil production.[111] In the opening paper the Soviet Oil Minister, concentrating on the perennial problem of irrational success indicators, called for closer integration of effort on the part of the many organizations involved in the productive process. He cited the instance of waterflooding of oil deposits* as an irrational indicator, in that performance is judged and rewarded on the basis of volume injected, which led to excessive flooding in a number of fields, causing additional problems in preparing oil for refining. This indicator has been abolished for the tenth Plan, the sole criterion being the total amount of oil produced.[112] Several papers in this symposium made reference to the delays caused by inadequate infrastructural development,[113] and the recurrent problem of equipment availability.[114]

It is important to put these issues into perspective. Russell has rightly pointed out that complaints in the oil industry were as much a feature of the sixties, when the oil industry regularly fulfilled its targets, as of the seventies. In his view the difference in the seventies has been that this criticism came from the industry's most senior officials and from the Oil Minister himself.[115] The concern

*the injection of water into the oil-bearing layer under pressure in order to facilitate extraction as natural pressure in the well declines.

at so high a level probably stemmed from the fact of higher expenditures in development and growing technical difficulties in production, encountered in an area as complex logistically as West Siberia. Whereas in the sixties oilfields were comparatively favourably located in relation to consuming centers and to each other, which meant a straightforward expansion of the logistic systems, bringing new productive capacity on stream in the seventies often involved the design and installation of substantial lengths of new pipeline and associated facilities. West Siberia has had to overfulfil its original targets during the ninth Plan to compensate for failure to meet targets elsewhere. Taking this into account the final Soviet production figure for 1975 of 491 million tonnes of oil and gas condensate compared with an original target of 505 million represents, at 97.2 per cent fulfilment, a considerable achievement.

The Soviet Gas Industry 1971-1975: Objectives and Achievements

The industry's objective for the ninth Plan was to raise production of natural gas from 198 billion cubic metres in 1970 to 320 billion in 1975, primarily through the development of gasfields in Tyumen' (West Siberia), Turkmenia, the Komi ASSR and the Orenburg oblast'. It was also planned to reduce wastage of well-head gas associated with oil production, with the objective of securing utilization of 85 to 87 per cent of such gas by 1975.[116] A total growth figure of 122 billion cubic metres was scheduled, of which West Siberia was to produce 35 billion, Turkmenia 52, the Komi ASSR 9 and the Orenburg oblast' 25.[117] The production balance in 1975 compared with 1970 was intended to change in accordance with data given in table 3.15.

Information on the extent and distribution of Soviet gas reserves is a good deal more abundant than it is in the case of oil, and gas distribution by location and depth is outlined in table 3.16. It has proved difficult, however, for Soviet geologists to gauge accurately the extent of gas reserves since the industry has developed comparatively recently and with great speed. Many of the newly discovered fields have proved among the largest in the world, as further geological information on each field resulted in revisions of original reserve estimates.

Table 3.15 Planned Change in Soviet Gas Output by Producing Area, 1975 Plan compared with 1970 (billion cubic metres)

	1970		*1975 Plan*	
	Production	*% of Total*	*Production*	*% of Total*
Total USSR	198	100	320	100
of which: European				
Russia and Urals	139	70.2	164.1	51.3
Orenburg obl.	1.3	0.7	26	8.1
Komi ASSR	6.9	3.5	16.1	5
Regions East of Urals	59	29.8	155.9	48.7
of which:				
West Siberia	9.3	4.7	44	13.8
Turkmenistan	13.1	6.6	65.1	20.3

Source: *Pyatiletnii plan 1971-1975*, pp 106-107

Table 3.16 Distribution of Total Gas Reserves in the USSR (%) (1971 data)

	Distribution by Region (% of total USSR)			Distribution by Depth (% of regional total)		
	Total	*At depth of 5,000 metres*	*At depth of 5,000–7,000 metres*	*Total*	*At depth of 5,000 metres*	*At depth of 5,000–7,000 metres*
USSR	100	100	100	100	90	10
of which:						
North-West	5.6	5.8	4.0	100	93	7
Urals	3.7	3.4	6.4	100	83	17
Volga	5.0	3.6	17.1	100	65	35
North Caucasus	2.3	1.6	9.0	100	59	41
Ukraine	4.1	2.8	15.0	100	61	39
Transcaucasia	0.8	0.5	4.0	100	50	50
West Siberia	32.5	36.1	–	100	100	–
East Siberia	13.9	15.4	2.1	100	98	2
Far East	14.2	15.8	–	100	100	–
Central Asia	10.7	10.0	17.1	100	83	17
Kazakhstan	7.2	5.0	25.3	100	64	36
European Russia plus Urals	21.5	17.7	55.5	100	57	43
Eastern Regions	78.5	82.3	44.5	100	94	6

Source: A E Probst, Ya E Mazover (eds), *op cit*, p 52

Authoritative estimates of total reserves in categories A+B+C1 were raised from 9.4 trillion (10^{12}) cubic metres in 1969[118] to 15.7 trillion at the beginning of 1971.[119] By 1975 a senior specialist of Gosplan, Yu I Bokserman, indicated that reserves in these categories had increased in the period from 1969 to 1974 by 13.1 trillion cubic metres to a total of 22.5 trillion.[120] These data are given in table 3.17. During the eighth Plan the rate of discovery of new gas reserves in West Siberia, Turkmenia, the Komi ASSR and the Orenburg oblast', though reported to be impressive, was insufficient to compensate for a shortfall in discovery in areas close to consumption centers, where targets for reserve discovery were underfulfilled.[121]

The rate of appreciation of reserves in the latter part of the eighth Plan and earlier part of the ninth was higher than the rate of growth in production.[122] By 1972 it was established that Siberian reserves in all categories extended to 12.6 trillion cubic metres, and those of European Russia and Central Asia to 3.7 trillion each.[123] Although this indicated that the medium- and long-term development of the gas industry depended on the utilization of Siberian resources, there was not the urgency as in the oil industry to bring Siberian operations on stream, since the ratio of reserves to production in European Russia was still favourable. However, at the outset of the ninth Plan it was feared that explored reserves in European Russia were being rapidly exhausted. Approximately 70 to 80 per cent of reserves discovered prior to 1971 had been brought into production and though European Russia was likely to maintain production up to 1975, its capacity to do so beyond that time was considered to be limited.[124]

Table 3.17 Accumulation of Soviet Natural Gas Reserves in Categories A+B+C1 1965-1974 (trillion cubic metres)

	1965	*1969*	*1970*	*1974*
Total USSR	3,220	9,423	12,100	22,500
of which: European Russia	1,771	2,152	2,583	4,400
West Siberia	315	5,146	7,116	14,100
East Siberia and Far East	91	377	439	700
Central Asia and Kazakhstan	1,043	1,747	1,962	3,300

Sources: Yu I Bokserman, 'Razvitie gazovoi promyshlennosti SSSR', *Neftyanik*, 2/1970, p 8; Yu I Bokserman, 'Nekotorye tendentsii dal'neishego razvitiya toplivnoi promyshlennosti', *Neftyanik*, 1/1975, p 5; Yu I Bokserman, 'Puti povysheniya effektivnosti transporta topliva', *Planovoe khozyaistvo*, 2/1975, p 21

During the ninth Plan activity in the exploration sector was characterized by the need to drill deeper in order to discover new reserves. Moreover in three of the areas designated to be of prime importance in the ninth Plan, Tyumen', the Komi ASSR and the Orenburg oblast', the incidence of complex, multi-component gas mixtures became more common, necessitating the installation of refining and separation facilities at the production center.[125]

At the outset of the ninth Plan it was estimated that Siberia contained some 60 per cent of the Soviet Union's discovered reserves of gas, the largest amount located in West Siberia.[126] By 1974 it was established that explored reserves in the four principal gasfields, namely Medvezh'e, Zapolyarnoe, Urengoi and Yamburg, totalled 9 trillion cubic metres.[127] The authoritative estimate made by V S Bulatov of the Tyumen' Gas Research Institute, of the amount of gas that could be extracted from the Tyumen' fields, put reserves at 40 trillion cubic metres. Within this figure, estimated reserves in 'industrial' categories totalled 11.8 trillion cubic metres, believed to be one quarter of explored reserves throughout the world, and three fifths of the Soviet total. Bulatov indicated that the optimal annual rate of production should be 5 to 6 per cent of reserves in the industrial categories, which amounts to 600 to 700 billion cubic metres in the case of the Tyumen' oblast.[128] The growth in proved gas reserves per metre of exploratory drilling in the Tyumen' oblast' has proved eight times greater than the all-Union average. Whereas only 9 per cent of exploratory drilling was carried out in the area during the period of the eighth Plan, it accounted for 70 per cent of the increase in discovery of reserves.[129] Some 75 per cent of West Siberia's proved reserves of gas are concentrated into the four fields mentioned above and this allows for the concentration of production for some 15 to 20 years.[130] However, Bulatov's estimate of possible production levels was based on the assumption that commensurate processing and transportation facilities would become available. Probst and Mazover, took into account the problems experienced in developing the region and suggested a likely peak annual production level of 350 to 450 billion cubic metres.[131]

In 1970 the production potential of the Medvezh'e field was estimated at

100 billion cubic metres per year.[132] Production commenced in 1972 at a level of 1.5 billion cubic metres, rising to 8 billion in 1973, destined mainly for the Serov metallurgical combine and industrial plants of the Sverdlovsk area.[133] The most advanced technology available has been directed to the Medvezh'e field with the result that it has been possible to drill wells of greater than average diameter. By 1974 the production rate was 1.5 to 2 million cubic metres of gas per day compared with the all-Union average of 200 thousand.[134] As a result of development completed during the ninth Plan it is considered that the Northern gas-bearing area of the Tyumen' oblast' should contribute 90 to 95 per cent of planned growth in Soviet gas production.[135] However, the need to bring the Urengoi field on stream in 1978 was acknowledged to be of primary importance for the continuing development of the gas industry in West Siberia.[136] During the ninth Plan production in the Tyumen' oblast' grew from 9.3 billion cubic metres in 1970 to 35.5 billion in 1975.[137]

The gasfields of Central Asia and Kazakhstan were scheduled to reach peak production during the ninth Plan, and it was considered that this could be sustained for a few years.[138] It is the Soviet Union's oldest producing area, and the major supplier of gas to the industrialized central zone. Given the existence of a high-capacity delivery system from the Central Asian Economic Region to the Centre Region, exploratory work was intensified in the neighbouring areas of Turkmenia and Uzbekistan to provide for the anticipated decline in production in the more developed areas. The Turkmen ASSR contains some 30 fields, of which the largest are the Shatlyk, Naip, Mary and Gugurtli gasfields and the Achak gas and gas condensate field. The Achak field was the first to be discovered in the Turkmen republic and its development dates from 1966. Uzbekistan possesses the huge Gazli field, development of which commenced even earlier, in 1959.

The contribution of the Turkmen gas industry has been a striking feature of the expansion of Soviet gas production during the ninth Plan. By 1970 only two out of 20 gas deposits discovered, namely the Achak and Maikop fields, were in production, with a potential output in that year of 11.6 billion cubic metres, or 88.5 per cent of Turkmen production.[139] Five years later it was estimated that production would be more than four times the 1970 level due to the development of newly discovered fields in the Eastern part of Turkmenia. In the event these accounted for the bulk of Turkmengazprom's 1975 production of 47 billion cubic metres.[140]

The major contributor was the Shatlyk field, which is believed to contain a total of 1,500 billion cubic metres of recoverable gas.[141] This field was brought into production in 1974, at which time Soviet officials predicted a production level of 35 billion cubic metres for 1975, based on extraction from 11 wells,[142] though this figure was eventually revised downwards to 32 billion.[143] The Naip field had the advantages of convenient location (near the trunk pipeline connecting Central Asia to the Centre Region, which enabled the gas to be fed under natural pressure directly into the delivery system), and an absence of sulphurous and carbonic impurities. These factors prompted the decision to develop the field rapidly, increasing production from 2 billion metres in 1972 to 15 billion in 1975.[144] The Naip field is believed to contain 1,800 billion cubic metres of recoverable gas.[145]

Gas production in Uzbekistan reached 36.2 billion cubic metres in 1974, at which time it was estimated that recoverable reserves ran to 1,180 billion cubic metres.[146] Continuous exploratory work in Western Kazakhstan during the ninth Plan resulted in the discovery of the Zapadno-Teplovskii gasfield, situated close to the main oil pipeline from Mangyshlak that joins the central distribution network in the Volga region.[147]

Though gas production in European Russia was scheduled to rise during the ninth Plan, its share of total production was to decline from 70.3 per cent in 1970 to 51.3 per cent in 1975.[148] The prime centers of interest were the Orenburg oblast' and the Komi ASSR. At the outset of the ninth Plan a 1975 production level for the Orenburg oblast' of 25 to 30 billion cubic metres of gas was foreseen.[149] This estimate was revised to 24 billion a year later,[150] and the actual level achieved in 1975 was 19.6 billion.[151] The particular value of Orenburg condensate is that in addition to providing methane for use as fuel, it is also possible to derive ethane, sulphur compounds and heavy hydrocarbons for use as feedstock in the chemical and petrochemical industries, which are priority sectors for local development.[152] However, the first section of the production and processing complex was completed some 18 months behind schedule in July 1974.[153] The delays experienced in commissioning new capacity in the Orenburg oblast' contributed to the shortfall in gas production in the area. It appeared that the impact of these delays would affect the first part of the tenth Plan, since the second stage of development, timed for completion in August 1975, and intended to increase the productive capacity of the area by 25 per cent, was well behind schedule by the early part of that year.[154] The importance of the area can be gauged from the fact that levels of local industrial development alone, based on fuel and feedstock from this deposit, suggest a potential consumption of 60 billion cubic metres.[155] The discovery of the Vuktyl gas condensate field during the eighth Plan marked the beginning of the contribution of the Komi ASSR to Soviet gas production. It was anticipated that in the medium-term the area would become second in importance to West Siberia as a gas-producing region. Production increased from 6.8 billion cubic metres in 1970 to 18.5 billion in 1975,[156] with total production between 1971 and 1975 of 77 billion.[157]

In the older producing areas, such as the Ukraine, the Volgograd oblast' and the Stavropol krai, slight increases or decreases were recorded during the ninth Plan, and their declining share of Soviet production is expected to continue. It should be noted that the Ukraine was still the biggest single producing area during the ninth Plan, although its share in total gas output declined from 30.8 per cent in 1970 to 22.3 in 1975. Its significance lies in the fact that until the completion of the pipeline from the Orenburg field it will remain the principal supplier of gas for export and for industrial consumers in the European zones of the Soviet Union.[158]

The contribution of individual areas to total Soviet production in 1975, compared with 1970, is detailed in table 3.18, and the performance of the production associations in table 3.19.

An area of activity which has grown in importance during the ninth Plan is that of the production and utilization of wellhead gas, found jointly with crude

Table 3.18 Soviet Gas Production by Area 1970 and 1975

	1970	*1975*
Total USSR	197.9	289.3
(%)	(100.0)	(100.0)
of which:		
Krasnodar krai	16.42	11.37
% of total	8.3	4.0
Orenburg oblast'	1.31	20.07
% of total	0.66	7.0
Tyumen' oblast'	9.28	35.50
% of total	4.7	12.3
Komi ASSR	6.80	18.50
% of total	3.4	6.4
Saratov oblast'	3.41	1.02
% of total	1.7	0.3
Volgograd oblast'	4.02	2.92
% of total	2.0	1.0
Ukraine	60.87	68.20
% of total	31.0	23.6
Turkmenistan	13.11	52.34
% of total	6.6	18.1
Uzbekistan	32.10	37.13
% of total	16.2	13.0
Azerbaidzhan	5.52	9.26
% of total	3.0	3.2

Source: R D Margulov, E K Selikhova, I Ya Furman, *Razvitie gazovoi promyshlennosti i analiz tekhniko-ekonomicheskikh pokazatelei*, Moscow, Ministerstvo Gazovoi Promyshlennosti, 1976, p 6

Table 3.19 Soviet Gas Production by Major Association 1970 and 1975 (billion cubic metres)

		1970	*1975*	*1975 as % of 1970*
Total USSR		197.9	289.3	146
of which:	Kubangazprom	22.5	5.8	25
	Stavropolgazprom	15.7	10.5	67
	Ukrgazprom	55.0	58.5	106
	Komigazprom	6.2	17.8	290
	Uzbekgazprom	31.5	36.6	120
	Turkmengazprom	11.8	47.0	400
	Orenburggazprom	0.8	18.4	X 22.2
	Tyumengazprom	9.2	33.5	360

Source: *ibid*, pp 6, 29-38

oil. At the outset of the ninth Plan there was a satisfactory rate of utilization of this gas in the older oil-producing areas, such as Azerbaidzhan, the Ukraine, Tataria and Bashkiria, varying from 85 to 94 per cent of production. However, in other areas, notably Turkmenia, the vast majority of wellhead gas was flared due to the lack of processing and transport facilities. In 1970 the Soviet oil industry flared some 14 billion cubic metres of wellhead gas.[159]

The directives for the ninth Plan included guidelines for utilization of wellhead gas. The percentage of such gas used in 1975 rose to 85-87 per cent of production compared with 61.1 per cent recorded in 1970.[160] Of the 22.9 billion cubic metres used in 1970, 11.4 billion was processed by the Ministry of the Gas Industry for fuel and 11.5 billion directed into the petrochemical industry as feedstock.[161] By 1972 an increase in utilization of 12.1 per cent above the 1970 level was recorded: however, it was in the period from 1974 to 1975 that utilization was planned to expand substantially when new capacity was to be commissioned.[162] At the end of the ninth Plan the use of wellhead gas showed an increase of 5 billion cubic metres above the 1970 level, reaching over 28 billion.[163] Six plants for processing this type of gas were constructed and commissioned between 1970 and 1975,[164] facilitating the development of the resources of Turkmenia and West Siberia. In the case of Turkmenia, the multistrata Kotur-Tepe field yielded over 300 cubic metres of wellhead gas per tonne of oil produced,[165] and by the end of the ninth Plan productive capacity had risen to 5 billion cubic metres per year.[166] West Siberian development is centered on the Nizhnevartovsk deposit where the available volume of wellhead gas is such that the expansion of the Tomsk and Tobol'sk petrochemical complexes has been planned on the basis of this gas as feedstock.[167] Though precise data on levels of utilization of wellhead gas achieved in relation to target during the ninth Plan have not been disclosed, it appears that in at least 10 oil-producing regions 80 per cent utilization was obtained.[168] In the Bashkir oilfields, which are at an advanced stage of exploitation, utilization reached over 90 per cent.[169]

During the ninth Plan exploratory work was carried out in West Siberia (North Tyumen'), East Siberia, offshore in the Baltic and Caspian Seas, and in Sakhalin. In view of the rapid discovery of gas reserves in the North Tyumen' area towards the end of the eighth Plan, the development of East Siberia is not critical in the short-term. Maintenance of exploratory effort has given rise to the Soviet claim that they possess the world's largest proven reserves of gas (40 per cent of total).[170]

By 1975 little work had been done in East Siberia. The first areas that were explored were the Vilyuisk region and the Yakut ASSR. Ultimate development of these gas reserves depends on decisions concerning the future level of industrialization and the concomitant demand for fuel. It also depends on the export opportunities primarily to Japan, via a possible pipeline to the Pacific coast at Nakhodka.[171] (The background to Soviet development of East Siberia, possibly with Japanese and American aid, is outlined in Appendix A.)

The total offshore gas-bearing potential area of the Soviet Union was estimated at 3 million square kilometres in 1970, with potential reserves of 10 trillion cubic metres off the Soviet Northern coast and 1 trillion in the Azov Sea.[172] Some minor offshore gasfields have been discovered in the Baltic and

Caspian Seas, though the prospects of discovering oil seem more promising.

Analysis of performance against original Plan in the gas industry reveals a substantial shortfall between 1971 and 1975. Problems similar to those in the oil industry have been encountered, namely the shortage and poor quality of equipment, high labour turnover in the less hospitable areas of exploration and production, and poor coordination between the production and distribution functions.[173] Table 3.20 outlines the growing divergence between Plan and performance from 1971 to 1975. It will be seen that the industry as a whole underfulfilled its original target in the second year of the Plan and subsequently. The overall effect was that of creating a gas deficit of 71 billion cubic metres, or 85 million tonnes of standard fuel, against the original Plan, as calculated in table 3.5.

Table 3.20 Growth Pattern in Soviet Gas Production 1971-1975 (billion cubic metres)

Year	*Plan*	*Actual*
1971	13.1	14.45
1972	18.0	8.98
1973	21.0	14.94
1974	30.0	24.23
1975	30.0	28.70
1970-1975	123.1	91.4
1970-1975 Annual average	24.6	18.3

Source: *Pyatiletnii plan 1971-1975*, p 346; Margulov, Selikhova, Furman, *op cit*, pp 5-6

There was a severe depletion of reserves in a number of producing areas, particularly the Ukraine and the North Caucasus. In addition production declined in some fields: the Krasnodar krai in particular showed a marked decline from 1971 to 1975. The problem of depletion was expected to increase in many of the Central Asian fields and this meant growing demands on available capital in order to maintain production levels in existing operations. It will be seen from table 3.21 that there was little change in the proportion of capital committed to offsetting depletion during the eighth and ninth Plans. The indication is that planners did not expect the provision for depletion to be so high towards the end of the ninth Plan. This compounded the difficulties anticipated in higher processing costs associated with the chemical characteristics of the gas produced in many of the more recently discovered fields.[174] A further important feature of development in the ninth Plan is that out of a total of 205.8 billion cubic metres' new capacity scheduled for completion, some 145 billion, or 70 per cent, was to be concentrated in six gasfields.[175] Delays experienced in bringing this new capacity on stream could not be counterbalanced elsewhere: capital resources were fully utilized since the investment Plan for the industry as a whole was fulfilled.

The inability of the gas industry to fulfil the original targets of the Plan was due not only to the technical problems encountered in the production process:

Table 3.21 Capital Utilization in the Soviet Gas Industry, 1966-1970 and 1971-1975 (million rubles)

Indicator	*1966-1970*	*1971-1975*
Capital Investment Total	1,019.1	3,139.4
to compensate for depletion in existing operations	502.2	1,591.0
to expand new production	516.9	1,548.4

Source: Margulov, Selikhova, Furman, *op cit*, p 21

there were also several difficulties in the provision of pipeline transport, necessary for the integration into existing systems of projected new production. Study of the logistic support sector sheds further light on the development of the oil and gas industries.

Developments in Transportation and Refining 1971-1975

The predominant method of transport used for Soviet oil and gas is the long-distance large-diameter pipeline. Table 3.22 outlines the growth in length and load turnover on the system from 1970 to 1975. There are significant cost advantages in transporting oil and gas by pipeline in preference to road, rail or, where possible, waterway. It was estimated at the beginning of the ninth Plan that transporting a given volume of oil and gas by pipeline was some two to three times cheaper on average than by rail, and one and a half to two times

Table 3.22 Installation and Utilization of Soviet Oil and Gas Pipelines 1970-1975

(a) Oil Pipelines

Year	*Total Length Installed at Year-End (thousand kms)*	*Deliveries of oil and products (million tonnes)*	*Load Turnover (million tonne-kms)*
1970	37.4	339.9	281.7
1971	41.0	352.5	328.5
1972	42.9	388.5	375.9
1973	47.2	421.4	439.4
1974	53.0	457.2	533.4
1975	56.9	497.6	665.8

(b) Gas Pipelines

	1970	*1971*	*1972*	*1973*	*1974*	*1975*
Length Installed at Year-End (thousand kms)	67.5	71.5	77.7	83.5	92.1	99.2
Gas Transmitted (billion cubic metres)	181.5	209.8	219.9	231.1	245.7	279.4

Source: *Narodnoe khozyaistvo SSSR 1975*, p 474

cheaper than by waterway.[176] The recoupment period for capital investment is shorter in the case of pipeline installation than in other forms of transport. This is achieved normally within three to four years of commissioning, and the operation of the pipeline system itself is a good deal less energy-intensive than the alternatives.[177] Throughout the sixties there was a trend in pipeline construction towards the greater use of large-diameter pipe: in 1960 the upper limit on pipeline diameter was 500 millimetres, by 1970 diameters of 1,020 and 1,200 millimetres were common.[178] The fact that Siberia became the Soviet Union's principal producing area in the course of the ninth Plan gave rise to a fundamental change in the pattern of interregional oil flows. As early as 1967 it was foreseen that by the end of the seventies the Western boundary of Siberia, which was then drawn in the Volga-Ural area, would become the dividing line between Westward and Eastward oil flows.[179] The area of distribution for West Siberian oil was seen to be Siberia itself, the Far East, North Kazakhstan, the northern part of European Russia and the eastern area of the Urals.

The importance of the development of the oil pipeline system can be judged from the fact that for the period 1971 to 1975 the level of capital investment was, at 3.9 billion rubles, three times that of the preceding Plan.[180] The rationale behind this decision to use the pipeline system related to the need to exploit Siberian resources. The West Siberian oil and gas province is twice as large as the Volga-Ural area; which is the nearest to it in size. Unlike the latter area, which contained an industrial waterway and rail system prior to the commencement of energy developments, Siberia was almost entirely virgin territory. Planners appreciated from the outset that the long-term significance of Siberia, for the economy as a whole, was greater than the Volga-Ural area had been. The population density was lower in Siberia than in the Volga-Ural area and the construction of a pipeline of given capacity considerably less labour-intensive than that of a rail system. On account of these factors, the harsh climate and difficulties in maintaining a regular supply of materials it was decided that, initially, only the minimum infrastructure adequate to sustain the oil and gas developments, would be undertaken.[181]

Progress in the construction of oil and gas pipelines was outlined in Elliot's (1974) work:[182] the state of development at the end of 1975 is detailed in tables 3.23 (oil) and 3.24 (gas). During the ninth Plan some 51.2 thousand kilometres of new trunk oil and gas pipeline were constructed in the Soviet Union, expanding capacity by 50 per cent. In the oil sector the following pipelines were constructed: the Usa-Moscow line (1,855 kms), the Samotlor-Al'met'-evsk line (2,313 kms), the line from Aleksandrovskoe to Irkutsk (1,766 kms), from Kuybyshev to Novorossiisk (1,522 kms) and from Nizhnevartovsk to Kuybyshev (1,183 kms). So too was the second string of the trans-Comecon 'Druzhba' pipeline, expanding substantially the Soviet Union's delivery capacity for Comecon and West European markets.[183] However, the original Plan called for an increase in installed length of 30 thousand kilometres;[184] hence at 19.5 thousand (table 3.23) performance was well short of the goal. In the gas sector the main lines added during the ninth Plan were over eight thousand kilometres in the system connecting Central Asia with Central Russia, over five thousand kilometres in the Medvezh'e-Central Russia system, approximately

Table 3.23 Basic Specifications of Major Oil Pipelines Completed between 1960 and 1975

Year of Commissioning	*From*	*To*	*Length (kms)*	*Diameter (inches)*	*Capacity (million tonnes/yr)*
1963	Al'met'evsk	Mozyr (i) Brest (ii) Uzhgorod	5,000	40/48	50
1964	Al'met'evsk	Irkutsk	3,700	28	20
1969	Baku	Batumi	883	8	na
1970	Uzen'	Kuybyshev	1,500	40	na
1973	Aleksandrovsk	Anzhero-Sudzhensk	820	40/48	na
1973	Anzhero-Sudzhensk	Krasnoyarsk	500	40/48	na
1974	Kuybyshev	Novorossiisk	1,520	32	na
1974	Samotlor	Al'met'evsk	2,130	48	na
1975	Usinsk	Yaroslavl'	1,800	28	na

Source: BEICIP/RRI, *The Petroleum Industry of the Soviet Union*, 1975, Part 2, p 115 (Geological/geographical consultancy report, made available by courtesy of The British Petroleum Company Ltd.)

2,500 kilometres in the Ukrainian system, over 800 kilometres of the Orenburg-Pskov line and approximately 600 kilometres as part of the second string of the line from Ukhta to Torzhok.[185] Performance in the gas sector showed a parallel, if less serious, underfulfilment of the original Plan in that the original target was to increase the installed length of gas pipeline by 60 per cent, to a total of 108 thousand kilometres.[186]

An additional problem that affected the underperformance of the gas-producing industry was that not all the increased capacity installed during the ninth Plan could be utilized due to the lack of equipment and control systems on the pipelines. Though the pipeline from Medvezh'e to Nadym had a capacity of 56 billion cubic metres per year, only 40 billion could be transmitted in 1975. Also in 1975 the delivery system from the Punga storage complex reached a capacity of 42 billion cubic metres; only 30 billion could be dispatched.[187] Delays in fully commissioning major pipelines not only limited delivery capacity but also gave rise to relative over-utilization of rail transport with consequent higher transport cost than planned. On average transport costs constitute 10 per cent of the final cost of industrial products, in the case of oil and petroleum products this figure is 25 per cent.[188] One method under discussion for lowering the transport cost element, delivery of oil saturated with gas by large-diameter pipeline, is thought to be readily applicable in areas where both oil and gas are produced in close proximity. The transport cost for oil and gas can be reduced by up to 20 per cent through using this method, but the scope for its potential application is limited by the fact that a point is quickly reached at which the increased energy cost incurred in operating the system exceeds the transport cost saving.[189]

The refining sector displayed a number of important developments during the ninth Plan. Though precise data on objectives are somewhat sparse, it is possible to discern the changing trends. It is significant that within the directives to

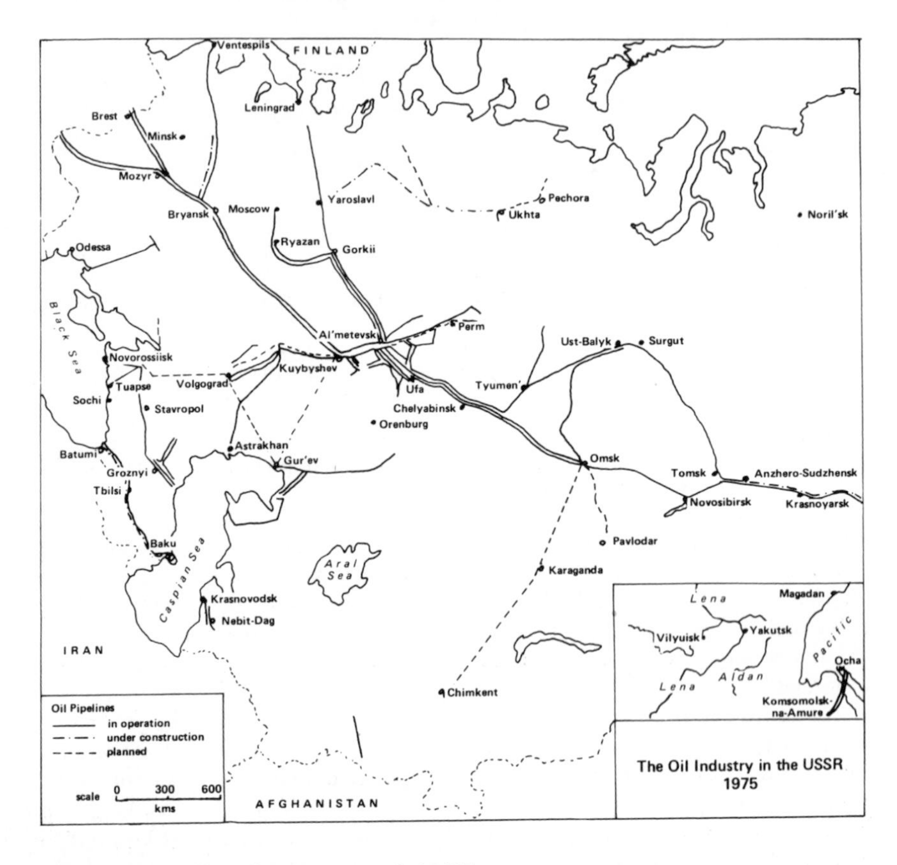

The Soviet Oil Industry at end-1975

Source: Adapted from J Bethkenhagen, *Bedeutung und Möglichkeiten des Ost-West-Handels mit Energierohstoffen* (Deutsches Institut für Wirtschaftsforschung, Sonderheft 104) Berlin, Duncker & Humblot, 1975, pp 62-63.

Table 3.24 Specifications of Major Gas Pipelines Completed between 1960 and 1975

Year of Commissioning	*From*	*To*	*Length*	*Diameter*	*Capacity (billion cubic metres/yr)*
1959/60	Shebelinka	Moscow	850	na	10.5
1962 (expanded 1973)	Bukhara	Sverdlovsk	4,973 (inc. branches)	40	19.5
1966	Berezova	Sverdlovsk	960	40	10.0
1969	Gazli	Moscow	3,740 (inc. branches)	40	13.0
1969	Gazli	Moscow (2nd line)	3,200 (inc. branches)	40/48/56	13.0
1970	Okarem	Ostrogozhsk, Voronezh	2,510	21/28/40	13.5
1970	Shekhitli	Ostrogozhsk	3,640	48/56	na
1970 (expanded 1975)	Bukhara	Alma-Ata	3,500	32/40	10.2 (1970) 25.0 (1975)
1970	Caucasus	Moscow	3,000	48/56	49.0
1972 (expanded 1975)	Medvezh'e	Moscow	3,000	48/56	38.0 (1972) 50.0 (1975)
1973	Shekhitli, Shebelinka	Moscow	3,000+	na	na
1973	Shebelinka	Dolina	1,115	40/48	12.3
1975	Medvezh'e	Brest	3,400	48/56	na

na = not available

Sources: BEICIP/RRI, *op cit*, (Part 2), p 112

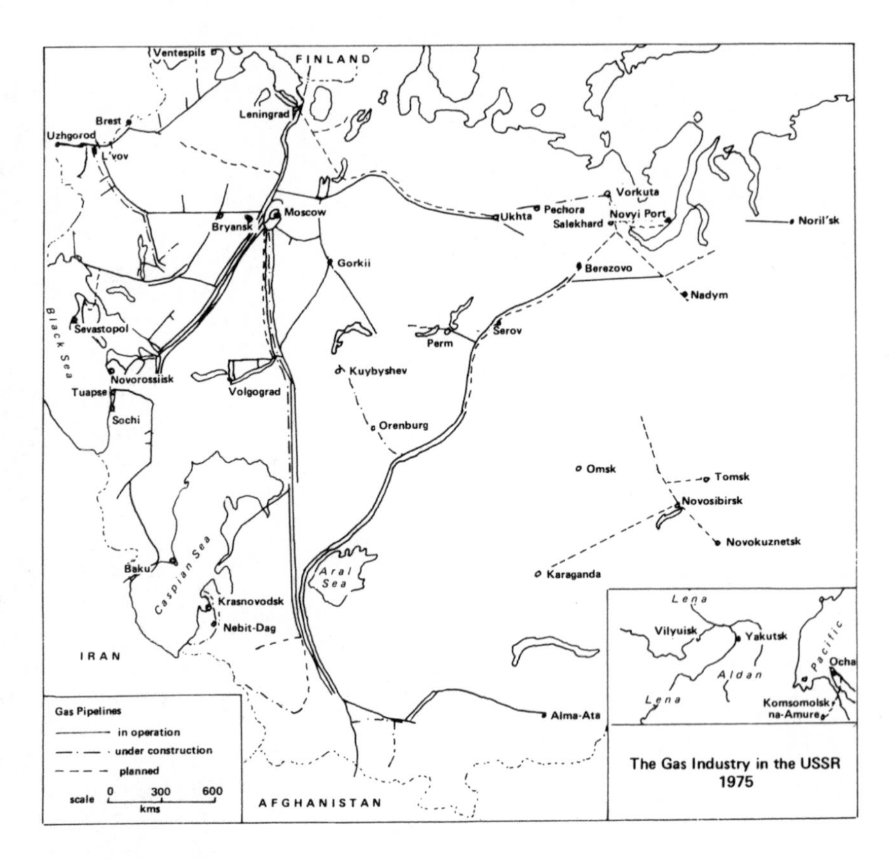

The Soviet Natural Gas Industry at end-1975

Source: Adapted from V A Smirnov, 'Gazovaya promyshlennost', *Ekonomika i organizatsiya promyshlennogo proizvodstva*, 5/1975, p 51

expand refining capacity, greater emphasis was placed on the manufacture of motor gasoline (the output of which in 1975 was 2.3 times that of 1970), on low-sulphur distillate fuel (scheduled to rise in 1975 to 1.4 times the 1970 level), and on winter grade diesel fuel (the output of which in 1975 was to rise to 1.75 times the 1970 level).[190]

European Russia is characterized by a high demand for motor gasoline and fuel oil in relation to other refined products and hence the refinery balance is geared to maximum production of these two product groups. In the Eastern regions the greater part of demand is for diesel fuel and petrochemical feedstock, and the refineries there seek minimum output of fuel oil. Natural gas and coal, which are available in the Eastern regions, are selected in preference to fuel oil. It is reported that in 1970 the refining balance of the Eastern regions showed little difference from that of European Russia. Relatively small quantities of light products were transported from European Russia to Central and Eastern areas, counterbalanced by deliveries of fuel oil from the latter area to the former.[191]

In accordance with the directives for the refining sector the refining balance as a whole altered during the ninth Plan in favour of maximum output of light (non-substitutable) products based increasingly on West Siberian crude oils, which are of low sulphur content and have a high yield of light products.[192] However, the underfulfilment of the Plan by the oil and gas industries caused a number of problems in oil refining. As a result of the substantial underfulfilment by the gas industry, and the inability of the coal industry to compensate by overfulfilment, the level of fuel oil that had to be produced during the ninth Plan was excessive, and caused some measure of under-supply of gasoline, distillate fuels and bitumen stock.[193] Furthermore, as a result of the suboptimal refining balance regional product supply was disrupted and cross-hauls of products were found to be necessary.[194]

Siberia was designated a priority area for the refining industry during the ninth Plan. Demand for all products and especially for gasoline and petrochemical feedstock was expected to rise. The output of these two product groups was to be maximized, using Kansk-Achinsk coal or Tyumen' gas locally rather than fuel oil as fuel.[195] The development of the Soviet refining industry has been characterized by a growing concentration of capacity in a small number of plants. Over the period from 1951 to 1960 the share in total output of refineries with a capacity of 1 million tonnes per year or less fell from 17.5 to 5 per cent. The process of production concentration was particularly intense in the period from 1961 to 1970, when the average capacity of a Soviet refinery increased more than twofold. Current development policy favours refineries of at least 6 million tonnes capacity per year. In 1972 these larger units accounted for approximately half the number of refineries and three quarters of refinery capacity in the Soviet Union.[196] In the case of Siberia the development of the refining industry was centered on the Omsk and Angarsk refineries. By the end of the ninth Plan it was considered that these refineries had reached their optimum size, beyond which any benefits of economy of scale in refining would be offset by additional costs of transportation and storage. At that stage a decision was taken to expand the Achinsk refinery and to draw up plans for further plants, timed to coincide with the expansion of oil production in West Siberia.[197]

Central Asia and Kazakhstan were faced with a substantial gap between productive capacity and consumption levels of refined products. By the end of 1975 the local refining sector could satisfy less than 25 per cent of demand in these areas.[198] However, the availability of gas has done much to alleviate this regional problem. European Russia has had adequate refining capacity throughout the post-war period and there has been an excess of capacity in the Volga-Ural area since the commencement of its development. The rate of expansion of capacity in European Russia during the ninth Plan was lower than in other regions and was confined to those areas experiencing a local product shortage.[199]

In the course of the ninth Plan the major problem affecting the refining industry was that of phasing in West Siberian crude oils. These oils were refined in Siberia, the Far East, Bashkiria, and the Kuybyshev oblast' and plans were drawn for West Siberian oils to be refined in Central and Southern Russia. The low sulphur and salt content of these oils makes them easier to refine than the standard Volga-Ural oils, and consequently puts them at a premium.[200] The value of these oils compounds the loss incurred in having to overproduce fuel oil. In order to meet a higher than anticipated demand a proportion of refined products had to be downgraded into fuel oil blending during the ninth Plan.

Financial Aspects of Soviet Oil and Gas Development 1971-1975

A study of pricing policy for crude oil and refined products sheds light on the Soviet objectives during the ninth Plan which were to optimize the use of oil in relation to coal and gas and to secure the desired movement of refined products to centers of consumption. Further objectives were the maintenance of the financial viability of individual enterprises and the narrowing of price differentials between purchasing enterprises.[201] Thus on 1 January 1971 the price of Tyumen' oil was cut by 18 per cent with the result that consumption increased as desired, and unit production costs decreased. At the same time the price of Bashkir crude oil was raised in order to offset increased production costs caused by reserve depletion and the consequent need to extract from deeper deposits.[202] The pricing policy for refined products has now been reappraised. Whereas up to the late sixties prices of all refined products were determined on the basis of a calorific comparison with other fuels, it was decided that such a policy would apply only in the case of fuel oil. Price levels of naphtha, middle distillates, and the crude oils designated for maximum output of these two product groups would reflect a desired competitiveness, both as feedstocks and fuels, against the alternatives, natural gas and gas liquids (ethane and propane).[203]

The pricing of wellhead gas illustrates the Soviet attempt to integrate the pricing mechanism with the policy of securing an optimal fuel balance. Despite directives for the ninth Plan to increase the utilization of wellhead gas, no more than 60 per cent of available gas was utilized due to the lack of processing capacity.[204] The bottleneck was caused by the fact that the prices for raw wellhead gas and its derivatives made investment recoupment well outside the norm, and financial returns during the build-up of productive capacity unattractive. Moreover production costs for wellhead gas were expected to rise in all areas except Tyumen'. It was suggested therefore that production costs should be reckoned separately for oil and wellhead gas, implying that the latter carried unrealistically high costs, 'subsidized' oil production, and that wholesale prices for wellhead gas be lowered as a temporary adjusting measure to stimulate demand. Such prices could then be readjusted in relation to oil and natural gas once the desired equilibrium had been attained.[205]

However, the pricing mechanism must reflect, above all, the rising average cost of oil production. Even the rapid development of Siberia, where production costs in 1972 were some 29 per cent below the all-Union average, could not compensate for the upward trend recorded in 20 out of 26 oil-producing areas.[206] The oil and gas industry recognizes that, in the long-term the problem of the rising prime cost of production will become more severe. This upward trend in the prime cost of oil and gas production compared with other raw materials is detailed in table 3.25. As will be seen, coal is the only raw material to exhibit falling production costs after the 1967 price reform. Since 1973 the position of coal has been reversed as production has moved increasingly to more difficult terrain.

Table 3.25 Changes in the Prime Cost of Selected Raw Materials (2nd half of 1967 = 100)

Material	*1965*	*1970*	*1973*
Iron Ore	89	106	116
Manganese Ore	92	132	125
Bauxite	78	101	111
Coal	92	97	96
Oil (inc. gas condensate)	74	105	112
Natural Gas	35	115	127
Apatite Concentrate	116	109	114
Natural Sulphur	81	102	114
Phosphorus	99	98	119

Source: Yu Yakovets 'Dvizhenie tsen mineral'nogo syr'ya, *Planovoe khozyaistvo*, 6/1975, p 4

The oil industry is relatively capital-intensive. Capital allocated to the industry totalled (in 1973 values) 4.05 billion rubles during 1951-1955; by 1961-1965 this figure had increased to 6.7 billion and during the ninth Plan over 14 billion rubles were allocated for oil development. This figure was close to 10 per cent of total industrial investment.[207]

Although the amount of capital allocated to the oil industry during the ninth Plan increased by 60 per cent above the level of the eighth, the rate of growth of production and reserve appreciation declined substantially.[208] Moreover it seems that in the latter part of the ninth Plan the original capital allocation had to be increased. By mid-1975 it was reported that the planned investment level for the oil industry had been raised to 15.5 billion rubles and that the expected uptake by the end of the Plan could be as high as 16.7 billion.[209] In the ninth Plan more detailed information was available on the structure of capital investment in the gas industry than it was in oil, and this is shown in table 3.26. The level of capital investment required during this period to secure an additional unit of production

Table 3.26 Structure of Capital Investment in the Soviet Gas Industry, 1970 and 1971-75 (million rubles)

	1970	*1975*	*1971-1975*
Gas Ministry total	1,189.61	2,966.9	10,903.5
%	100	100	100
Geological exploration	10.9	78.0	335.0
%	0.9	2.6	3.1
Production	332.7	878.5	3,139.4
%	28.0	29.6	28.8
including: production drilling	114.9	170.62	701.9
gasfield construction	217.8	707.9	2,437.5
transportation	695.78	1,718.45	6,445.8
%	58.5	57.9	59.1
Processing	36.73	102.77	478.0
%	3.1	3.5	4.4
Underground Storage	7.2	62.04	116.9
%	0.6	2.1	1.0
Machine Manufacture	22.7	57.2	131.4
%	1.9	1.9	1.2
Other	83.6	69.92	257
%	7.0	2.4	2.4

Source: Margulov, Selikhova, Furman, *op cit*, p 17

and processing capacity increased by factors of 2.6 and 2.0 respectively.[210] Table 3.27 shows the pattern of capital investment in 1975 compared with 1970, and table 3.28 the overall trend in investment in the fuels industries for the whole of the ninth Plan.

The Soviet Oil and Gas Industries 1971-1975: An Overview

In the course of the ninth Plan the rate of growth in the Soviet oil and gas industries showed a decline that was characteristic of many industrial sectors of the economy. In the case of the oil industry, only the intensive development of the new, remotely located fields could stave off a serious shortfall. However, the accelerated rate of extraction gave rise to a greater need to prove new reserves in the short-term, and this, in itself, proved a problem area. But the under-performance of the economy as a whole generated a lower energy demand than was envisaged.

The general trend in oil and gas was towards a greater tension in the balance of supply and consumption, complicated by increasing difficulties in distribution. These were associated with the accelerated development of West Siberian resources. Natural gas became the most rapidly growing fuel within the energy balance despite the industry's poor performance in 1972 and the consequent underfulfilment of the original annual targets to the end of the ninth Plan. The overriding priorities of energy policy for the tenth Plan were the need to conserve conventional fuel resources, especially oil, and to accelerate the development of nuclear power.

Table 3.27 Analysis of Growth in Capital Requirement of Soviet Gas Industry (All figures are rubles per 1,000 cubic metres)

	1970	*1966-1970 Annual Average*	*1975*	*1971-1975 Annual Average*	*Average 1971-1975 as % of Average 1966-1970*	*1975 as % of 1970*
Capital Investment for 1,000 cubic metres' new capacity	12.2	7.81	13.96	15.3	196	114
Capital Investment for 1,000 cubic metres' production growth	22.3	15.8	35.7	40.8	258	160
of which: production drilling	7.71	5.1	6.93	9.1	178	90
Construction cost for 1,000 cubic metres gas extracted	2.64	1.56	3.78	3.25	208	143

Source: Margulov, Selikhova, Furman, *op cit*, p 19

Table 3.28 Capital Investment in the Soviet Fuels Industries, 1970 and 1971-1975 (million rubles, constant prices)

	1970	*1971*	*1972*	*1973*	*1974*	*1975*	*1971-1975 Total*
Industry Total	28,597	30,275	32,400	34,112	36,630	39,852	173,269
%	100	100	100	100	100	100	100
Oil Industry	2,527	2,759	3,003	3,080	3,491	3,853	16,186
%	8.8	9.1	9.3	9.0	9.5	9.7	9.3
Gas Industry	1,041	1,122	1,229	1,482	1,738	1,798	7,369
%	3.6	3.7	3.8	4.3	4.7	4.5	4.3
Coal Industry	1,541	1,624	1,713	1,743	1,728	1,759	8,577
%	5.4	5.4	5.3	5.1	4.7	4.4	5.0

Source: *Narodnoe khozyaistvo SSSR 1975*, p 508

References

1. A E Probst, 'Puti razvitiya toplivnogo khozyaistva SSSR', *Planovoe khozyaistvo*, 6/1971, p 54.

2. *ibid.*

3. *ibid.*

4. *ibid*; also *Gosudarstvennyi pyatiletnii plan razvitiya narodnogo khozyaistva SSSR na 1971-1975 gody*, Moscow, Politika, 1972, p 97. (This is referred to hereafter as *Pyatiletnii plan 1971-1975*.)

5. *Pyatiletnii plan 1971-1975*, p 103.

6. Probst, *op cit*, p 54; *Pyatiletnii plan 1971-1975*, p 107.

7. *Pyatiletnii plan 1971-1975*, p 108.

8. *ibid.*

9. *ibid.*

10. *ibid*, pp 108-109; Economist Intelligence Unit, *Soviet Oil to 1980*, London, EIU, 1973, p 37.

11. A Vigdorchik et al, 'Metody optimizatsii dolgosrochnogo razvitiya toplivno-energeticheskogo kompleksa SSSR', *Planovoe khozyaistvo*, 2/1975, p 30.

12. *Sotsialisticheskaya industriya*, 29 September 1973, p 2; P P Galonskii, 'Bol'shie perspektivy, vazhnye zadachi', *Neftyanik*, 2/1972, p 2.

13. M A Styrikovich, 'Naucho-tekhnicheskie problemy razvitiya energetiki SSSR', *Izvestiya AN SSSR, Ser. energetika i transport*, 3/1974, p 8.

14. A Mel'nikov, V Shelest, 'Toplivno-energeticheskii kompleks SSSR', *Planovoe khozyaistvo*, 2/1975, pp 10-11.

15. Editorial 'Uskorenno razvivat' toplivno-energeticheskuyu bazu strany', *Planovoe khozyaistvo*, 2/1975, p 5. The Ministries of the Chemical and Petrochemical and Metallurgy industries are criticized for not having advanced this policy.

16. Probst, *op cit*, p 58.

17. L A Melent'ev, A A Makarov, 'Osobennosti optimizatsii razvitiya toplivno-energeticheskogo kompleksa', *Izvestiya AN SSSR, Ser. energetika i transport*, 3/1974, p 12.

18. *Pyatiletnii plan 1971-1975*, p 98.

19. *ibid*, p 99.

20. *ibid.*

21. M I Rostovtsev, T G Runova, *Dobyvayushchaya promyshlennost' SSSR*, Moscow, Mysl', 1972, p 16.

22. Mel'nikov, Shelest, *op cit*, p 13.

23. A S Pavlenko, A M Nekrasova (eds), *Energetika SSSR v 1971-1975 godakh*, Moscow, Energiya, 1972, p 31.

24. *ibid*, p 32.

25. Murray Feshbach (Rapporteur), *Report of a visit by delegation of US systems analysts to Moscow*, 1974. Mimeo, pp 5-6.

26. *ibid*, pp 6-7. Note also that research into the decision-making process of the Soviet energy sector as a whole is being carried out by D Wilson and Dr P Hanson of the Centre for Russian and East European Studies, University of Birmingham, UK.

27. *ibid*, p 5.

28. *ibid*, p 7.

29. W Gumpel, 'Energiebilanzen als Mittel der Energieplanung in der UdSSR', *Jahrbuch der Wirtschaft Osteuropas*, Munich, 1972, Volume 2, pp 295-317.

30. *ibid*, p 296.

31. Melent'ev, Makarov, *op cit*, p 13. A particularly pressing problem is that of forward assumptions of prices in relation to costs, this allegedly posing difficulties in deriving an optimal balance.

32. *ibid*, p 15.

33. *ibid*.

34. Styrikovich, *op cit*, p 8.

35. The advantages in conversion efficiency that arise from the use of natural gas are outlined in I F Elliot, *The Soviet Energy Balance*, New York, Praeger, 1974, pp 236-243.

36. E Yudin, 'Effektivnoe ispol'zovanie toplivno-energeticheskikh resursov strany – obshchegosudarstvennaya zadacha', *Planovoe khozyaistvo*, 6/1975, p 60.

37. *ibid*, p 63.

38. V V Mikhailov, 'Problemy ekonomiki promyshlennogo energopotrebleniya', *Promyshlennya energetika*, 1/1976, p 9.

39. Elliot, *op cit*, p 263.

40. See esp *Planovoe khozyaistvo*, 2/1975, p 7; and Yu I Bokserman, 'Nekotorye tendentsii dal'neishego razvitiya toplivnoi promyshlennosti', *Neftyanik*, 1/1975, p 5.

41. R W Campbell, 'Some Issues in Soviet Energy Policy for the Seventies', in *Soviet Economic Prospects for the Seventies*, Washington DC, US Congress, Joint Economic Committee, 1973, p 55.

42. Jeremy Russell, *Energy as a Factor in Soviet Foreign Policy*, Farnborough, Saxon House, 1976, p 200.

43. P Galonskii, 'Neftyanaya, gazovaya i neftepererabatyvayushchaya promyshlennost' v 1971-1975 godakh', *Planovoe khozyaistvo*, 11/1972, p 31.

44. *ibid*, p 32.

45. Elliot, *op cit*, p 80; Russell, *op cit*, p 40; J Bethkenhagen, *Bedeutung und Möglichkeiten des Ost-West-Handels mit Energierohstoffen*, (Deutches Institut für Wirtschaftsforschung, Sonderheft 104), Berlin, Duncker & Humblot, 1975, p 55; R W Campbell, *The Economics of Soviet Oil and Gas*, Baltimore, Johns Hopkins Press, 1968, p 68. (A definition of the Soviet method of classifying reserves is given in appendix C.)

46. *Petroleum Press Service*, August 1969, p 307.

47. R E Ebel, 'Two Decades of Soviet Oil and Gas', *World Petroleum*, vol 42, No 5, pp 78, 84.

48. *Neftyanoe khozyaistvo*, 1/1969, p 2.

49. Of total reserves in categories A+B+C1, 15 per cent lie at less than 1,200 metres, 66.4 per cent at 1,201-1,800 metres, 11.2 per cent at 1,801-2,400 metres, 5.2 per cent at 2,401-3,000 metres and 4.3 per cent at more than 3,000 metres. Elliot, *op cit*, pp 81-82.

50. *ibid*, p 82.

51. A E Probst, Ya A Mazover (eds), *Razvitie i razmeshchenie toplivnoi promyshlennosti*, Moscow, Nedra, 1975, p 82.

52. *ibid.*

53. B P Orlov, 'O ratsional'nykh metodakh osvoeniya prirodnykh zapasov v novykh raionakh Sibiri', *Ekonomika i organizatsiya promyshlennogo proizvodstva*, 2/1971, p 26.

54. The production cost in the giant Samotlor oilfield is as low as 50 per cent of the all-Union average. See V P Maksimov, 'Perspektivy i zadachi razvitiya proizvoditel'nykh sil Tyumenskoi oblasti', *Ekonomika i organizatsiya promyshlennogo proizvodstva*, 2/1972, p 38.

55. The late Soviet Oil Minister V D Shashin notes in 'Puti povysheniya effektivnosti neftyanoi promshlennosti', *Planovoe khozyaistvo*, 4/1973, p 21, that operations in the important northern area of the Tyumen' oblast' have accounted for only 15 per cent of total drilling in West Siberia. He also notes that over the recent (unspecified) period not a single new oilfield has been opened in the Tomsk oblast'.

56. Orlov, *op cit*, p 26.

57. F G Arzhanov et al, 'Nekotorye prakticheskie resheniya po avtomatizatsii operativnogo upravleniya neftyanoi promyshlennost'yu Zapadnoi Sibiri', *Neftyanoe khozyaistvo*, 6/1974, p 2.

58. Probst and Mazover (eds), *op cit*, pp 86-87.

59. N Mal'tsev, 'Neftyanaya promyshlennost' strany v desyatoi pyatiletke', *Neftyanik*, 2/1976, p 1.

60. E A Ogorodnov, 'Osvoenie mestorozhdenii nefti i gaza v Zapadnoi Sibiri', *Stroitel'stvo truboprovodov*, 3/1971, p 27.

61. *Izvestiya*, 14 December 1975, p 6.

62. M A Mkhchiyan et al, 'Nefti sredneobskoi neftegazonosnoi oblasti Zapadnoi Sibiri', *Khimiya i tekhnologiya topliv i masel*, 4/1974, p 3.

63. *ibid.*

64. V Bogachev, 'K voprusu ob intensifikatsii osvoeniya prirodnykh bogatstv Sibiri', *Kommunist*, 3/1973, pp 89-100.

65. *ibid*, p 91.

66. *ibid*, p 92.

67. *ibid*, p 99. Bogachev argues as follows: 'Was it possible to fix precisely the costs of exploratory drilling in the lower Ob' area for 1971-1975? All exploration is uncertain and hence it is impossible to calculate forward costs even in promising areas with good indications of possible extraction levels.'

68. I Ognev, 'Postizhenie otkrytiya', *Ekonomika i organizatsiya promyshlennogo proizvodstva*, 4/1976, p 174.

69. *ibid*, p 175.

70. *ibid*, p 176. Note that the full report of the conference proceedings was published under the title *Neft' i gaz Tyumeni dokumentakh*, Sverdlovsk, 1973.

71. *Pyatiletnii plan 1971-1975*, p 258; *Pravda*, 1 February 1976, p 2.

72. F G Arzhanov, Yu B Fain, 'Samotlor – unikal'noe neftyanoe mestorozhdenie', *Neftyanoe khozyaistvo*, 5/1975, pp 38-44.

73. R I Kuzovatkin, 'Front i tyl Samotlora', *Ekonomika i organizatsiya promyshlennogo proizvodstva*, 6/1976, pp 78-87.

74. It has been proved possible to sink up to 16 wells from a single platform. See Arzhanov, Fain, *op cit*, p 141, and Kuzovatkin, *op cit*, p 80.

75. Arzhanov, Fain, *op cit*, p 44. Also *Pravda*, 11 June 1975, p 3.

76. Probst, Mazover (eds), *op cit*, p 90.

77. *ibid*, p 91.

78. It is reported that Turkmenneft' and Republican geological organizations had failed to fulfil annual plans for reserve discovery 'for years'. *Turkmenskaya iskra*, 23 May 1970, p 1.

79. *ibid*, 5 September 1971, p 2.

80. *Pravda*, 20 October 1972, p 1.

81. *Kazakhstanskaya pravda*, 31 August 1973, p 2.

82. Editorial 'Nekotorye itogi neftedobyvayushchei promyshlennosti v 1969 godu i zadachi na 1970 god', *Neftyanoe khozyaistvo*, 2/1970, p 2.

83. *Vyshka*, 11 June 1974, p 1.

84. V I Kudinov, 'Segodnya i zavtra Udmurtskoi nefti', *Neftyanoe khozyaistvo*, 7/1975, p 4.

85. N A Mal'tsev, 'Permskaya oblast'', *Neftyanoe khozyaistvo*, 3/1971, pp 33-35.

86. S Fedorchenko, 'Na puti k novym rubezham', *Neftyanik*, 6/1976, p 5.

87. To facilitate extraction it is sometimes necessary to maintain pressure in the oil-bearing layer by injecting steam or nitrogen. This process is difficult and adds to operating costs. Discussed in P D Alekseev, 'Orenburgneft' v devyatoi pyatiletke', *Neftyanoe khozyaistvo*, 1/1976, p 7.

88. Probst, Mazover (eds), *op cit*, p 87.

89. Z V Driatskaya et al, 'Nefti Komi ASSR', *Khimiya i tekhnologiya topliv i masel*, 4/1975, p 7.

90. N Kochurin, 'Chtoby vypolnit' zadaniya Partii', *Neftyanik*, 5/1976, p 6.

91. *ibid.*

92. For a brief outline of Belorussian oil developments see Elliot, *op cit*, p 106.

93. G Topuridze, B Golenishchev, 'U neftyanikov Poles'ya', *Neftyanik*, 6/1976, p 8.

94. This is discussed in N Zhmykova et al, 'Nefti mestorozhdenii Pribaltiki', *Neftyanik*, 6/1976, p 19.

95. A Namestnikovs, V Bergmanis, 'Kadas ir Baltijas dziles?', *Cina*, 19 December 1974, p 4.

96. *Vyshka*, 26 September 1973, p 2.

97. *ibid*, 22 February 1975, p 2.

98. B N Zykin, N E Podkletnov, 'O ratsional'nykh sposobakh ispol'zovaniya Sakhalinskoi nefti', *Izvestiya SO AN SSSR Ser. obshch. nauk.*, 6/1970, pp 109, 111.

99. *ibid*, p 109.

100. Probst, Mazover (eds), *op cit*, p 91.

101. A brief history of the development of East Siberia is given in Elliot, *op cit*, p 106. An earlier work, V Conolly, *East Siberian Oil*, Mizan, August 1971, pp 16-21, gives a little more detail.

102. In 1960 East Siberia was estimated to contain over half the hydroelectric potential of the Soviet Union and about 80 per cent of coal reserves. *Trudy konferentsii po razvitiyu proizvoditel'nykh sil Vostochnoi Sibiri*, Moscow, AN SSSR 1960, p 6.

103. A E Probst (ed), *Razvitie toplivnoi bazy raionov SSSR*, Moscow, Nedra, 1968, p 175.

104. K M Zvyagintseva, 'O trekh zonakh toplivno-energeticheskogo khozyaistva Sibiri', *Izvestiya SO AN SSSR Ser. obshch. nauk*, 1/1974, pp 20-21.

105. *Trud*, 24 October 1975, p 2.

106. A Babakuliev, 'O sovershenstvovanii planirovaniya i material'nom stimulirovanii v geologorazvedochnykh predpriyatiyakh', *Izvestiya AN Turkmenskoi SSR, Ser. obshch. nauk.*, 5/1969, p 52.

107. K K Smirnov, 'Dlya blaga rodiny', *Stroitel'stvo truboprovodov*, 11/1974, pp 2-3.

108. Labour turnover in Mangyshlak was 93 per cent in 1970 due allegedly to poor working and living conditions. *Narodnoe khozyaistvo Kazakhstana*, 10/1971, pp 45-49.

109. R N North, 'Soviet Northern Development: The Case of North-West Siberia', *Soviet Studies*, October 1972, p 188.

110. Workers are granted a one-off bonus of four times the average monthly wage in addition to having a regional wage coefficient of 1.7 times the average in European Russia. Rates of bonus rewards for fulfilment and overfulfilment of the Plan are above the national average. These bonuses are enhanced as basic salaries and wages are increased in accordance with length of service. Paid leave is extended to one and a half months per year with one free return ticket to any point in the Soviet Union for each worker and his family. Data presented in S A Orudjev, V I Muravlenko, 'Integrated Planning for Exploration, Development, Production and Transportation for Rapid Expansion of Oil Field Operations', *Proceedings of the 9th World Petroleum Congress*, Barking: Applied Science Publishers, 1976, Vol 3, p 336. Note that North (*op cit*, p 188) points out that the bulk of labour turnover is experienced not in petroleum operations but in support activity, where the aforementioned benefits are not applicable.

111. The conference was attended by senior officials of the CPSU Central Committee, Ministry of the Oil Industry, Ministry of the Gas Industry, Ministry of Oil and Gas Construction Enterprises, Trades Unions, et al. The main papers are compiled in *Ekonomika neftyanoi promyshlennosti*, 7/1975.

112. V D Shashin, 'Povyshenie effektivnosti neftyanogo proizvodstva – glavnoe napravlenie razvitiya otrasli', *Ekonomika neftyanoi promyshlennosti*, 7/1975, p 6.

113. For example, V I Muravlenko, in 'O problemakh razvitiya neftyanoi promyshlennosti v Zapadnoi Sibiri', *Ekonomika neftyanoi promyshlennosti*, 7/1975, p 13. He adds that even if the authorities were successful in attracting an adequate number of people to work in West Siberia the provision of accommodation was insufficient.

114. A A Asan-Nuri, 'O nekotorykh problemakh tekhnicheskogo progressa v oblasti burovykh rabot', *Ekonomika neftyanoi promyshlennosti*, 7/1975, p 19.

115. Russell, *op cit*, p 51.

116. *Pyatiletnii plan 1971-1975*, p 106.

117. *ibid*, pp 106-107.

118. Yu I Bokserman, 'Razvitie gazovoi promyshlennosti SSSR', *Neftyanik*, 2/1970, p 8.

119. Probst, Mazover (eds), *op cit*, p 49. See also A K Kortunov, 'Uspekhi gazovoi promyshlennosti i perspektivy ee razvitiya', *Gazovaya promyshlennost'*, 3/1971, p 1.

120. Yu I Bokserman, 'Puti povysheniya effektivnosti transporta topliva', *Planovoe khozyaistvo*, 2/1975, p 21.

121. Galonskii, *op cit*, p 33.

122. A K Kortunov, 'Gazovaya promyshlennost' Sovetskogo Soyuza', *Gazovaya promyshlennost'*, 8/1970, p 2.

123. V E Orel, V P Stupakov, 'Ratsional'no ispol'zovat' zapasy gaza', *Gazovaya promyshlennost'*, 3/1974, p 31.

124. Probst, Mazover (eds), *op cit*, p 50.

125. *ibid*, p 53. In the Orenburg deposit, for example, ethane, propane and other heavy hydrocarbons are found along with methane gas.

126. E N Altunin, 'Gaz i gazosnabzhenie Sibiri', *Stroitel'stvo truboprovodov*, 9/1970, p 6.

127. V D Chernyshov et al, 'Zadachi osvoeniya gazovykh mestorozhdenii na severe Tyumenskoi oblasti', *Stroitel'stvo truboprovodov*, 6/1974, p 12.

128. V S Bulatov, 'Analiz faktorov, opredelyayushchikh uroven' dobychi prirodnogo gaza na Tyumenskom severe', *Izvestiya SO AN SSSR Ser. obshch. nauk.*, 1/1974, p 12.

129. *ibid*, p 13.

130. *ibid*, p 14.

131. Probst, Mazover (eds), *op cit*, pp 60-61.

132. Altunin, *op cit*, p 6.

133. R Saifullin, 'Medvezh'e – gazovyi promysel', *Gazovaya promyshlennost'*, 11/1975, p 24.

134. E N Altunin, 'Osnovnye voprosy bystreishego osvoeniya mestorozhdenii Zapadnoi Sibiri', *Gazovaya promyshlennost'*, 8/1974, p 11.

135. *Ekonomicheskaya gazeta*, 19/1975, p 6.

136. V S Bulatov, 'Puti povysheniya effektivnosti gazodobychi na Tyumenskov severe', *Izvestiya SO AN SSSR Ser. obshch. nauk*, 6/1975, p 32.

137. R D Margulov, E K Selikhova, I Ya Furman, *Razvitie gazovoi promyshlennosti i analiz tekhniko-ekonomicheskikh pokazatelei*, Moscow: Ministerstvo Gazovoi Promyshlennosti 1976, pp 6, 38.

138. Probst, Mazover (eds), *op cit*, p 61.

139. Report by V Fedorov of a conference on the development of the oil, gas and petrochemical industries of Turkmenia, *Gazovaya promyshlennost'*, 9/1970, p 49.

140. A A Annaliev, 'Turkmengazprom na novom etape razvitiya otrasli', *Gazovaya promyshlennost'*, 3/1976, p 10.

141. *Turkmenskaya iskra*, 29 May 1971, p 1; *Kazakhstanskaya pravda*, 26 March 1972, p 1.

142. *Turkmenskaya iskra*, 1 February 1972, p 2.

143. *ibid*, 3 September 1972, p 2.

144. *ibid*, 18 July 1974, p 1.

145. P O Tarakanov, 'Osvaivaya gazovye bogatstva Turkmenii', *Stroitel'stvo truboprovodov*, 4/1975, p 12.

146. Editorial 'Zadachi razvitiya gazovoi promyshlennosti Uzbekistana', *Gazovaya promyshlennost'*, 2/1974, p 2.

147. *Kazakhstanskaya pravda*, 15 January 1974, p 2.

148. Galonskii, *op cit*, p 34.

149. G D Margulov, 'Orenburgskii gazopromyshlennyi kompleks – vazhnaya stroika devyatoi pyatiletki', *Stroitel'stvo truboprovodov*, 3/1971, p 37.

150. *Stroitel'stvo truboprovodov*, 4/1972, pp 5-6.

151. Margulov, Selikhova, Furman, *op cit*, p 35.

152. Yu V Zaitsev, 'Orenburgskii gazovyi kompleks i ego perspektivy', *Gazovaya promyshlennost'*, 3/1975, p 6.

153. *Vyshka*, 11 July 1974, p 1.

154. Report of a conference held on 29 February 1975 concerning the development of the Orenburg oblast', ('Orenburgskii gazovyi kompleks: itogi, problemy, resheniya'), *Gazovaya promyshlennosti'*, 5/1975, p 53.

155. Probst, Mazover (eds), *op cit*, p 62.

156. Margulov, Selikhova, Furman, *op cit*, p 6.

157. Kochurin, *op cit*, p 6.

158. V M Kuzenko et al, 'Sostoyanie ispol'zovaniya proizvodstvennykh fondov v dobyche gaza i puti ego uluchsheniya (ob"edinenie 'Ukrgazprom')', *Gazovaya promyshlennost'*, 12/1975, p 7.

159. Galonskii, *op cit*, p 34.

160. *Pyatiletnii plan 1971-1975*, p 106. In 1970 production of wellhead gas was 37.5 billion cubic metres, utilization 22.9 billion. See also V I Baraz, 'Sostoyanie i osnovnye napravleniya v ispol'zovanii neftyanogo gaza', *Neftepromyslovoe delo*, 10/1973, p 3.

161. Baraz, *op cit*, p 3.

162. *ibid*, p 4.

163. V I Baraz, 'Sostoyanie i osnovnye napravleniya uluchsheniya ispol'zovaniya neftyanogo gaza', *Neftepromyslovoe delo*, 7/1976, p 52.

164. These were: (i) Groznyi, capacity 3.1 billion cubic metres per year; (ii) Gnedintsev, 0.42; (iii) Kazakh, 1.0; (iv) Fifth unit of Minnibaev plant, 0.585; (v) Nizhnevartovsk, plant No 1, 2.0; (vi) Pravdinsk, 0.5. *ibid*, p 52.

165. P Tarakanov, 'Ispol'zovanie neftyanogo poputnogo gaza na mestorozhdenii Kotur-Tepe', *Neftyanik*, 3/1975, p 12.

166. Baraz, *op cit* (1976), p 52.

167. G V Tukov et al, 'O resursakh gaza v nefti Samotlorskogo mestorozhdeniya', *Neftyanoe khozyaistvo*, 2/1975, p 60.

168. Baraz, *op cit* (1973), p 2.

169. *Ekonomicheskaya gazeta*, 43/1975, p 16.

170. Editorial, 'Prognozirovanie perspektiv gazonosnosti', *Gazovaya promyshlennost'*, 11/1974, p 2.

171. E Dmitriev, 'Razvitie geologorazvedochnykh rabot na neft'i gaz v Vostochnoi Sibiri', *Neftyanik*, 4/1975, p 11.

172. S I Levin, 'Trebovaniya k podgotovke stroitel'stva morskikh truboprovodov', *Stroitel'stvo truboprovodov*, 12/1973, p 4.

173. Russell, *op cit*, pp 62-66, catalogues the problems faced by the gas industry during the ninth Plan.

174. Margulov, Selikhova, Furman, *op cit*, pp 20-21.

175. These were Medvezh'e, Shatlyk, Achak, Naip, Vuktyl and Orenburg. *ibid*, p 21.

176. Z S Prutyanova, 'Tekhniko-ekonomicheskie pokazeteli razvitiya truboprovodnogo transporta', *Ekonomika neftyanoi promyshlennosti*, 12/1970, p 38.

177. *ibid*, pp 38-39.

178. *ibid*, p 39.

179. V S Varlamov, 'Problemy transportnogo osvoeniya Zapadno-Sibirskoi nizmennosti v svyazi s formirovaniem na ee territorii novogo narodnokhozyaistvennogo kompleksa', *Izvestiya AN SSSR Ser. geog.*, 3/1967, p 52.

180. *Pyatiletnii plan 1971-1975*, p 103.

181. Varlamov, *op cit*, p 54.

182. Elliot, *op cit*, pp 55-59 and pp 109-117.

183. K K Smirnov, 'Vo imya mogushchestva Velikoi Rodiny', *Stroitel'stvo truboprovodov*, 2/1976, p 1.

184. *Pyatiletnii plan 1971-1975*, pp 103, 216.

185. Smirnov, *Vo imya . . .* (1976), p 1.

186. *Pyatiletnii plan 1971-1975*, p 204.

187. These and other shortfalls in delivery performance against design capacity are outlined in Margulov, Selikhova, Furman, *op cit*, pp 45-46.

188. Z S Prutyanova, 'Rezervy v nefteprovodnom transporte i metodika ikh opredeleniya', *Ekonomika neftyanoi promyshlennosti*, 3/1972, p 24.

189. This issue is discussed, mainly from the technical standpoint, in P N Uskov, 'Sovershenstvovanie odnotrubnogo sbora nefti i gaza', *Neftepromyslovoe delo*, 1/1972, pp 31-35.

190. Galonskii, *op cit*, p 36.

191. Probst, Mazover (eds), *op cit*, p 151.

192. The characteristics of West Siberian crude oils are given in M A Mkhchiyan et al, 'Nefti Zapadnoi Sibiri', *Neftyanik*, 12/1975, pp 17-18.

193. Editorial, 'Neftepererabatyvayushchaya i neftekhimicheskaya promyshlennost' v 1975 godu', *Khimiya i tekhnologiya topliv i masel*, 1/1975, p 6.

194. Probst, Mazover (eds), *op cit*, p 151.

195. *ibid*, p 152.

196. *ibid*, p 167.

197. *ibid*, p 157.

198. *ibid*, pp 157-158.

199. *ibid*, pp 158-159.

200. Editorial, *Khimiya i tekhnologiya topliv i masel*, 1/1975, p 5.

201. V K Vasil'eva, 'O tsenakh n neft', *Ekonomika neftyanoi promyshlennosti*, 11/1971, p 22.

202. *ibid.*

203. V K Vasil'eva, 'Sovershenstvovanie tsenoobrazovaniya v otrasli', *Ekonomika neftyanoi promyshlennosti*, 11/1974, p 58.

204. V K Vasil'eva, N N Kosinov, A Z Kuz'min, 'Nekotorye voprosy formirovaniya optovykh tsen na neftyanoi gaz', *Ekonomika neftyanoi promyshlennosti*, 2/1976, p 7.

205. *ibid*, pp 7-8.

206. V K Vasil'eva, 'Sovershenstvovanie tsenoobrazovaniya . . .', 1974, p 59.

207. G D Sokolov, *Kapital'noe stroitel'stvo v neftyanoi promyshlennosti*, Moscow, Nedra, 1973, p 3; *Narodnoe khozyaistvo SSSR, 1975*, p 508.

208. G D Sokolov, 'K voprosu otsenki effektivnosti kapital'nykh vlozhenii v neftyanuyu promyshlennost'', *Ekonomika neftyanoi promyshlennosti*, 7/1974, p 3.

209. A M Lalayants, 'O povyshenii effektivnosti proizvodstva i ispol'zovanii kapital'nykh vlozhenii v neftyanoi promyshlennosti', *Ekonomika neftyanoi promyshlennosti*, 7/1975, p 31.

210. Margulov, Selikhova, Furman, *op cit*, p 18.

Chapter 4

The Oil and Gas Industries of Eastern Europe to 1975

There is a marked difference in the exploration for and the production and utilization of fuel resources in Eastern Europe compared with the Soviet Union. Some countries are well provided with reserves of one or more fuels in relation to domestic needs, others are poorly endowed. This situation would give rise to differing priorities in energy policy, were each member country of the Council for Mutual Economic Assistance (Comecon) to pursue independent lines of development. However, a major element of the 1971 Complex Program of Economic Development is that of working towards an integrated energy policy optimal to the bloc as a whole. Central to this policy is the objective of securing a shift away from a predominantly coal-based energy economy and towards a greater utilization of oil and gas.

In Eastern Europe the trend towards oil and gas commenced in the mid-sixties. This trend was scheduled to gain momentum and to continue into the seventies.[1] In 1970, the share held by oil and gas in the consumption balance of Eastern Europe was as follows: the GDR 15.1 per cent, Poland 14.9 per cent, Czechoslovakia 20.5 per cent, Hungary 43.7 per cent and Bulgaria 45.8 per cent. Only in the case of Romania, the sole Eastern European country to possess substantial reserves of oil and gas, was the major share accounted for by hydrocarbon fuels, the corresponding figure being approximately 72 per cent.[2]

It should be stressed that in the sixties the Eastern European members of Comecon were not facing the problem of a net energy shortage: the problem then was one of securing an improved fuel and energy balance. This strategy would involve all countries except Romania becoming dependent on external sources of oil and eventually of gas. An additional problem was that the development of industries using hydrocarbon feedstock was beginning to generate stronger links between the economies of the Eastern European countries, though the extent of this differed in each case. In the process of coordinating development plans in the energy sector, there arose issues such as the extending of financial assistance by one country to another, optimal size and location of energy conversion plants of varying types and, not least, the nature of production specialization in fuels within the bloc in the movement towards integration.[3] The essence of the aforementioned issues is that an energy policy aimed at integrated planning of the development of disparate and widely dispersed resources involves shifts in macroeconomic and geopolitical relations within the bloc, particularly when the preferred resources are as flexible as oil and gas.

In the early sixties the debate on problems of energy provision in Eastern

Europe for the period from 1960 to 1980 centered on an estimated fivefold increase in the aggregate GNP of member-countries and a sixfold increase in the output of their industrial sectors.[4] The pattern of energy consumption in Eastern Europe from 1950 to 1970, together with an estimate for 1980, is given in table 4.1. Trends towards oil and gas were planned to differ in individual countries. In the case of Hungary and Bulgaria estimates made in early 1973 foresaw an oil and gas share in the total energy balance of 60 to 65 per cent by 1980, possibly rising to 75 per cent by the year 2000. In the GDR, Czechoslovakia and Poland the share of oil and gas would be significantly lower, in the range of 30 to 38 per cent by 1980, given the indigenous reserves of brown coal in the GDR and Czechoslovakia and hard coal in Poland.[5]

The emergent role of oil and gas in the Eastern European energy balance during the sixties is shown in table 4.2, and the developing trend in fuel production in

Table 4.1 Consumption of Energy in Eastern Europe, 1950-1970 and Estimate for 1980 (million tonnes of standard fuel)

Country	*1950*	*1960*	*1970*	*1980 Estimate*	*1960 as % of 1950*	*1970 as % of 1960*	*1980 Estimate as % of 1970*
Bulgaria	2.2	9.7	27.7	55.5	441	286	200
Hungary	9.1	21.4	34.8	45.0	235	163	129
GDR	47.6	87.8	100.0	120	184	114	120
Poland	52.2	97.3	142.3	181-186	186	146	127-131
Romania	7.3	36.1	58.6	96-98	495	162	164-167
Czechoslovakia	37.3	61.9	90.6	112	166	146	124

Sources: (i) I D Kozlov, E K Shmakova, *Sotrudnichestvo stran-chlenov SEV v energetike*, Moscow, Nauka, 1973, p 125; (ii) A Alekseev, Yu Savenko, *Ekonomicheskaya integratsiya v razvitii toplivno-energeticheskikh otraslei stran-chlenov SEV*, Voprosy ekonomiki, 12/1971, pp 49-52

Table 4.2 Energy Balance of Eastern Europe 1960-1970 (%)

Fuel	*Year*	*Bulgaria*	*Hungary*	*GDR*	*Poland*	*Romania*	*Czechoslovakia*
Coal	1960	64.0	72.3	96.0	90.0	17.0	84.0
	1965	61.2	65.0	90.4	86.1	17.1	81.7
	1970	53.7	47.8	81.4	82.0	18.7	78.2
Oil	1960	14.0	12.0	3.0	4.0	29.0	7.0
	1965	27.9	19.6	7.0	6.5	22.0	11.8
	1970	43.3	29.7	14.9	9.7	22.0	18.3
Gas	1960	–	2.5	–	1.0	46.0	–
	1965	0.6	6.0	0.1	3.0	50.0	–
	1970	2.5	14.0	0.2	5.2	54.0	2.2
Other	1960	22.0	13.2	1.0	4.1	8.0	9.0
	1965	10.3	9.4	2.5	4.4	10.9	6.5
	1970	0.5	8.5	2.5	3.1	5.3	1.3

Source: H Ufer, 'Wachstums- und Strukturprobleme der Energiewirtschaft der RGW-Länder', *Wirtschaftswissenschaft*, August 1975, p 1128

Table 4.3 Production of Major Fuels in Eastern Europe 1960-1970

	1960	*1965*	*1970*	*1970 as % of 1960*
Hard Coal (million tonnes)				
Bulgaria	0.6	0.6	0.4	67
Hungary	2.8	4.4	4.2	150
GDR	2.7	2.2	1.0	37
Poland	104.4	118.8	140.0	134
Romania	3.4	4.7	6.4	188
Czechoslovakia	26.4	27.8	28.2	107
Brown Coal (million tonnes)				
Bulgaria	15.4	24.5	29.0	188
Hungary	23.7	27.1	24.0	101
GDR	225.5	250.8	261.0	116
Poland	9.3	22.6	33.0	355
Romania	3.4	5.6	14.0	412
Czechoslovakia	57.9	72.3	81.0	140
Oil (million tonnes)				
Bulgaria	0.2	0.22	0.33	165
Hungary	1.2	1.80	1.84	153
GDR	–	–	–	–
Poland	0.19	0.33	0.42	220
Romania	11.5	12.5	13.37	116
Czechoslovakia	0.14	0.18	0.20	143
Natural Gas (billion cubic metres)				
Bulgaria	–	0.7	0.47	–
Hungary	0.34	1.1	3.48	X 10
GDR	–	–	–	–
Poland	0.55	1.3	4.97	X 9
Romania	9.80	16.7	23.9	244
Czechoslovakia	1.44	0.9	1.14	79

Source: Kozlov, Shmakova, *op cit*, pp 36-37, 48

table 4.3. This trend towards the greater use of oil and gas was influenced by two factors. First, it became evident during the sixties that the more thermally efficient hydrocarbon fuels could be supplied to a consumer at a lower delivered cost in standard fuel units than coal, fuelwood or other energy materials. This gave rise to a natural process of substitution. Decisions were made to fuel new productive capacity with oil and gas, and in certain cases to invest in the conversion of existing coal-burning plant to these fuels. Second, the growth of industries needing oil and gas as a feedstock compounded the substitution process, the major sectors being road transport, air freight and the petrochemical industry. This brought about a gradual change in growth rates in production and utilization of each energy source, and also a decline in net consumption of energy per unit of output.

It has been estimated that in 1970 total East European reserves of fuel in all categories were 150 billion tonnes of standard fuel, of which Poland possessed

67.5 per cent (101.2 billion), Czechoslovakia 11.6 per cent (17.4 billion), the GDR 8.7 per cent (13.1 billion), Romania 3.4 per cent (5.1 billion), Bulgaria 1.4 per cent (2.1 billion) and Hungary 1.2 per cent (1.8 billion).[6] It must be borne in mind that these figures relate to total reserves and hence include substantial quantities that will not be recoverable, also that the majority of the reserves consist of coal, a fuel less favoured than oil and gas.

At the outset of the 1971-1975 Plan Eastern Europe as a whole was known to be facing the problem of increasing dependence on imported energy though the degree of this dependence differed in each country. It was estimated that Bulgaria's energy demand would rise from 27.7 million tonnes of standard fuel in 1970 to 41.5 million in 1975, 55.5 million in 1980, 75 million in 1990 and 110-120 million in the year 2000. At the same time it was estimated that import dependence would rise from 60 per cent in 1960 to 74 per cent in 1980, a level which would be maintained until the end of the century. In Hungary it was expected that energy demand would rise from 34.8 million tonnes of standard fuel in 1970 to 45 million in 1980, the corresponding change in import dependence rising from 37 to 50 per cent. Energy demand in the GDR was a little over 100 million tonnes of standard fuel in 1970 and this was expected to rise to 120 million in 1980, with the possibility of reaching 140-150 million by the year 2000. It was estimated that Poland's energy demand would rise from 142.3 million tonnes of standard fuel in 1970 to 181-183 million in 1980, whereas in the case of Romania the corresponding rise would be from 78-80 million tonnes to 96-98 million, during which times its status as a net exporter of energy would be reversed to that of 80 per cent self-sufficiency. Czechoslovakia was expected to record growth in energy consumption from 90.6 million tonnes of standard fuel in 1970 to 112 million in 1980, with the likelihood that this would reach 149 million in 1990 and 205 million in the year 2000. It was estimated that Czechoslovakia's dependence on imported energy would rise from 20 per cent in 1970 to 40 per cent in 1980.[7]

The fundamental issue affecting energy planning for the bloc as a whole was the coincidence of Eastern Europe's move from coal to oil and gas with the increasingly difficult economic and logistic problems experienced by the Soviet Union, historically the bloc's largest supplier and preferred source.

In the early seventies Eastern European analysts and planners anticipated that the trend from consumption of solid fuels and hydroelectricity towards oil and gas would continue to 1980, along with a relatively minor contribution from nuclear power. However, the anomalous position of Romania, where the share of the energy balance held by oil and gas was expected to decline, calls for comment. The development of the oil and gas industries of Romania commenced at a relatively early stage with the result that by 1950 these two fuels accounted for some 75 per cent of Romanian energy consumption.[8] By 1970 the production growth rate of each fuel had declined, especially that of oil, whereas plans for the further rapid development of the economy still required an increasing supply of energy. It was decided that alternative energy sources would have to be exploited, including that of nuclear fuels. In addition, the domestic coal industry would have to maintain, and possibly expand, its share of the energy balance, and if necessary would be supplemented by increased imports.

Problems of Energy Planning in the Comecon Bloc

The formation of energy policy in Comecon is presented by Soviet and East European analysts as an example of integrated planning and production specialization: however, because of the very distribution of energy reserves the Soviet Union plays the leading role.

At the outset of the 1971-1975 Plan period the problems surrounding the provision of oil and gas in Eastern Europe were seen to be threefold. First, the inevitability of increased imports by Eastern Europe (Romania excepted) from the Soviet Union would have to be facilitated by the expansion of the trans-Comecon 'Druzhba' pipeline, involving the commissioning of a second line parallel to the original, the Northern branch of which would supply Poland, the GDR, the Southern branch Czechoslovakia and Hungary. Oil deliveries to Bulgaria would continue by tanker via the Black Sea. It was intended to increase the share of natural gas in the Eastern European energy balance by extending the 'Bratstvo' pipeline, through which Poland and Czechoslovakia had been supplied since 1967. Plans provided for the extension of this pipeline so that deliveries to the GDR via the trans-Polish branch, and to Hungary and Bulgaria via a new southern branch would commence in 1974.[9]

Second, the Soviet Union was viewed by the Eastern European countries as the prime agent in assisting them to develop such resources of oil and gas as they had, through the provision and exchange of technical expertise and, where appropriate, of basic production equipment.[10] In 1969 it was estimated by the East German analyst W Siegert that, taking into account trends in economic growth in Eastern Europe, the demand for Soviet oil in each of the Eastern European countries (Romania again excepted) would rise to twice the 1970 level by 1980.[11] A level of production in the Soviet Union of 600 to 700 million tonnes in 1980 was forecast, and aggregate demand in Eastern Europe minus Romania was put at 80 to 100 million tonnes.[12] The question of the role of gas exports and their substituting effect on the pattern of oil exports was not discussed in Siegert's analysis: however, at the time it was possible to detect a change in attitude on the Soviet part towards developing the gas industry rapidly, not only as a contributor to the domestic energy balance but also as an export fuel to complement oil and refined products, given the increasing difficulties of maintaining rates of growth in oil production.

The third problem was that of optimizing the investment pattern for the bloc's fuel industries. Allied with this was the extension of trading credits and appropriate bilateral arrangements to facilitate Soviet supply. The issue of investment is regarded as being a problem of scale and length of the recoupment period.[13] Investment plans in the fuel-producing countries are generally tightly balanced, with competition for resources from a number of enterprises.[14] From the Soviet viewpoint a function of joint investment is that of spreading the risk that inevitably accompanies exploration in the fuels industries, particularly in oil and gas: for the Eastern European countries the major consideration is the comparative efficiency of investment in assisting the expansion of the Soviet fuels industries, versus the immediate alternative of directing available funds into the development of domestic resources.

A later analysis advocated a three-dimensional approach to energy planning for the bloc as a whole. It was suggested that considerable scope existed for improving energy conversion efficiency; that recognition of the growing difficulties in energy production in the Soviet Union and its long-term implications might prompt the accelerated development of indigenous energy reserves; and that the energy-intensity of the Eastern European economies could be lowered by improvements in inter-branch planning so as to reduce the material-intensity of production and hence energy intensity.[15] By the end of the 1971-1975 Plan period it was admitted that attempts at securing energy autarchy would not solve the problem of energy provision: increased trade outside the bloc would have to be negotiated.[16]

From 9-12 December 1975 a symposium of energy economists from the member-countries of Comecon was convened in Moscow to discuss the above questions. The broad conclusions of the symposium were that total energy demand might be higher than previously estimated. Soviet experts indicated that an aggregate demand of 780 million tonnes of standard fuel in the full and associate member-countries, including Cuba, Mongolia and Yugoslavia but excluding the Soviet Union, was possible by 1980, and that this would rise to over 1 billion by 1990. They advanced the view that the level of self-sufficiency in the aforementioned group of countries would decline from 70 per cent in 1975 to 50 per cent by 1990 and that the import of energy resources of all types would increase threefold.[17]

The consensus of the symposium on the optimal pattern of development of the Comecon energy balance was that a rapid program of investment in nuclear power ought to be undertaken, but that even if such a plan were implemented it could not be expected to make a significant contribution before the eighties. Two elements of policy were highlighted as being particularly relevant to the short-term problem, namely that a target of 20 per cent improvement in conversion efficiency over 1975 levels could be achieved economically; and that this process could be supported by a more critical attitude as to the location and expansion of energy-intensive industry in relation to the existing and anticipated fuel supply.[18] It is significant that the prospects for increasing imports of oil from OPEC producers were considered limited and that a maximum of 30 million tonnes of OPEC oil at an estimated cost of 2 billion rubles could be contemplated by 1980.[19]

The impression gained from studying reports of Soviet and Eastern European discussions of energy questions is that planners and analysts admit the existence of a number of interacting problems, which while causing some concern have not reached the level of a Comecon 'energy crisis'. However, a different range of conclusions emerge from a number of Western analyses of energy development in Comecon, written in the late sixties and early seventies.

In a study written in 1969 the American analyst Jaroslav Polach, taking into account the changing trends in consumption of individual fuels and in general conversion efficiency, concluded that in 1980 Eastern Europe (in this instance including Albania and Yugoslavia) would face an energy deficit of 150 to 165 million tonnes of standard fuel, of which oil might account for 130 million, this being 100 million tonnes' actual oil production. If the Soviet Union were to supply this requirement it would account for 16 to 18 per cent of production in 1980, which was estimated by Polach at 550 to 600 million tonnes.[20]

In his view a supply of this quantity would seriously impair Soviet plans for economic growth. The alternative was increased imports from other oil producers and he maintained that this might hinder the overall process of economic integration within the bloc. In addition, these imports would need to be paid for partly in hard currencies which might necessitate the liberalization of trade between Eastern Europe and the rest of the world. This eventuality could work to the disadvantage of the Soviet Union in that it might then have to compete for the output of Eastern European industry. Polach advanced a further alternative – in his view the likely one – namely that the Soviet Union would seek to strengthen its hold on the fuel surplus of the Middle East as an interim measure, until the development of Siberia was further advanced, thus maintaining control of fuel exports to Eastern Europe without having to curtail domestic demand or exports to the West.[21]

In theory the Soviet Union might logically wish to pursue this course; however, Polach did not assess the feasibility of the course itself in the international context. At the time of his writing, the major oil companies, operating in various groupings in the Middle East, still had control over the decision-making process determining price and output of oil. There was a strong belief in the West that the prime objective of the Soviet Union's oil policy was to maintain as high a level of exports as possible in order to gain increased influence in the world market. Consequently, if the producing companies were to make substantial quantities of oil available to the Soviet Union, they would assist an export drive, possibly risking erosion of the market price in Western Europe. This had occurred during the Soviet re-entry into Western markets in the late fifties. Since the companies and producer governments were seeking to increase prices in Western Europe, the policy suggested by Polach would have worked to their disadvantage and would have been blocked.

In another article written in 1969 the American analyst Stanislaw Wasowski addressed himself to the question of the imminent oil deficit in Eastern Europe.[22] Wasowski based his projection of energy demand on a projection of an energy/GNP elasticity coefficient, including an adjustment for the assumed substitution effect of oil for solid fuels. A further assumption is that natural gas would not make any impact on the Eastern European energy balance. In the case of oil demand he analyzed the growth trend and planned development of those sectors of the different economies in which oil had no ready substitute, specifically citing transport and petrochemicals.* Wasowski's conclusion was that the demand for oil in Eastern Europe in 1980 could be expected to rise to 190 million tonnes.[23] He expressed the view that indigenous production of oil would be insignificant, that nuclear power could make no contribution by 1980 and that the solution would be found in Eastern Europe negotiating a supply on the basis of barter trade with Middle East producers.[24] However, the constraints which would apply in the case of supply to the Soviet Union would also affect

*Note, however, that the petrochemical industry had been designated a priority sector for the supply of gas.

Eastern Europe.

A different approach was taken by the German analyst Werner Bröll, writing at about the same time as the previous two authors.[25] By extrapolating an index of per capita energy consumption, and an unchanging percentage primary energy balance, he arrived at a similar conclusion to that reached by Wasowski, namely that Eastern Europe would face an oil deficit of 140 million tonnes in 1980. This represented a consumption level of 170 million tonnes less indigenous production of 30 million tonnes.[26] Bröll, alone in his time, qualified this estimate by stressing that the rapidly developing natural gas industry could alleviate the medium-term problems of energy supply, by substituting in certain end-uses where oil had hitherto been preferred. Equally important was Bröll's accurate anticipation of the need for Eastern Europe to invest directly in the development of Soviet hydrocarbon resources, with a payoff in contractually guaranteed deliveries.[27]

By 1973 it was evident that there were no signs of an increase in Eastern Europe's non-Soviet oil imports at a rate that might suggest the eventual levels predicted in earlier analyses. However, the prevailing view in the West was that Eastern Europe would nevertheless face an oil deficit. Writing in mid-1973,[28] Sabine Baufeldt took the view that previous analyses had overestimated energy demand, underestimated supply and ignored the emergent role of gas. Nonetheless she predicted an oil deficit of 100 million tonnes by 1980, putting demand at 120 million and indigenous supply at 20 million.[29] She also stressed that the capacity of Eastern Europe to increase indigenous production of oil was limited and that the short-term solution to the problem was to diversify the source of supply to include purchases from the Middle East.[30]

If, during the early seventies, Western perceptions of Eastern Europe's need to negotiate for supplies of oil from Middle East producers differed only in conclusions about the amount that would be involved, the events of 1973-1974 changed this perspective. Such was the theme of a Japanese analysis of the Eastern European energy problem written after the effect of OPEC's price rises had begun to be felt.[31] Kazuo Ogawa pointed to the benefits offered by the 'energy crisis' and the oil price rises to the energy- and chemical-exporting members of Comecon (the Soviet Union, Poland and Romania), and contrasted the situation of other members of the bloc.[32] He stressed that whereas the OPEC price rises afforded the Soviet Union the opportunity of raising the intra-Comecon prices of crude oil and refined products, the impact on Eastern Europe was to reverse the trend of 'weaning' from the Soviet Union. This process of 'weaning' had commenced in the late sixties and was expected in some quarters to continue. Now there was a greater urgency in Eastern Europe's drive to develop indigenous reserves.[33] Ogawa pointed to a further aspect of the impact of the OPEC price rises and the Soviet Union's difficulties in increasing oil deliveries, namely the re-evaluation and emergence of natural gas, which could become the Soviet Union's preferred export fuel for Eastern Europe.[34]

The analysis of the Canadian researchers B Korda and I Moravcik, published in 1976,[35] took the view that the problems of energy provision in Comecon were exacerbated by two essentially systemic features of the economic and political climate: the intrinsic disunity in the bloc,[36] and the propensity of the economic mechanism itself to prompt wastage of raw materials and energy.[37] They argued that Eastern Europe's energy resources, even if fully developed, would be insufficient to meet the countries' long-term needs; that the current energy mix was technically suboptimal; that the energy-intensity of production was high by international standards; and that lack of hard currency prevented the input of oil and gas from non-Comecon sources, which they regarded as the rational solution.[38]

They concluded that political expediency (which they did not define) was likely to remain the overriding factor influencing energy development in Eastern Europe, and that this was manifest in the Soviet Union's willingness to sustain economic loss through exporting to Eastern Europe rather than to world markets, in order to retain dominance in the political sphere. They foresaw five elements of strategy for dealing with the Eastern European problem: development of alternative energy forms (especially nuclear power), increased intra-Comecon trade in all forms of primary and secondary energy, increased inputs from outside the bloc (negotiated primarily by the Soviet Union), stringent energy conservation measures and a permanent rationing policy towards certain consumers.[39]

In order to analyze the contribution of the oil and gas industries of Eastern Europe to their countries' energy balance, account must be taken of the state of development of these industries at the end of 1975 and the immediate problems faced by planners at the outset of the 1976-1980 period.

Dispersion of Energy Resources in Eastern Europe

The fundamental features of the development of the Eastern European energy balance are that changes in fuel consumption to favour use of the more efficient fuels, oil and gas, have taken place in all these countries at a later stage than in the Soviet Union (Romania excepted) and that there has been a widening gulf between the consumption and production balance. By 1973 production of oil in Eastern Europe totalled 22.6 million tonnes, and of gas 49.5 billion cubic metres; these figures represented 1 and 4 per cent of world production respectively. In both the oil and gas industries in Eastern Europe there were considerable difficulties in proving reserves even though sizeable proportions of the land-mass are considered to be potential oil- and gas-bearing areas, as detailed in table 4.4.

During the sixties the rate of discovery of oil reserves in Hungary enabled planners to conclude that oil production levels could be maintained. In Poland, Czechoslovakia and the GDR the declining ratio of oil reserves to production prevented plans for expansion, and it was felt from the early seventies that Romania might also face this problem.[40] As in the case of the Soviet Union, data on reserves are sparse. Lisichkin gave no indication of the level of reserves in Poland, the GDR and Bulgaria; however, he put the level of proven oil reserves in Romania at 130 million tonnes,[41] and in the case of Czechoslovakia and Hungary confined himself to the statement that reserves are 'insignificant'.[42] Maksakovskii was more forthcoming with data on oil reserves, all of which related to the categories A+B+C1, but which were calculated from information compiled in 1967. He put Polish reserves at 8 million tonnes, those of the GDR at 1.5 million,

Table 4.4 Potential Hydrocarbon Bearing Area in Eastern Europe (thousand square kilometres)

Country	*Total Area*	*Potential Hydrocarbon Bearing Area*	*% of Total Area*
Poland	312.5	255	81.2
GDR	108.3	63	58.2
Czechoslovakia	127.9	36	28.1
Hungary	93.0	74	79.6
Romania	237.5	150	63.2
Bulgaria	110.9	48	43.3
Total	990.1	626	63.2

Source: V P Maksakovskii, *Toplivnye resursy sotsialisticheskikh stran Evropy*, Moscow, Nedra, 1968, p 21

of Czechoslovakia at 1.7 million, Romania at 100 million and Bulgaria at 2.1 million: his figure for Hungary was for total ultimate reserves, and he put this at 36.8 million tonnes.[43]

Lisichkin was also sparing in his data on gas reserves. He put the Hungarian total in the categories A+B+C1 at 115 billion cubic metres,[44] gave no data for Poland, Czechoslovakia, Romania and Bulgaria, and viewed any estimate of reserves in the GDR as premature due to the late commencement of exploration.[45]

For his part Maksakovskii, again relying on 1967 data, put Polish gas reserves in the aforementioned categories at 200 billion cubic metres, those of Czechoslovakia at 13 billion, with the likelihood of a further 13 billion in the category C2, of Hungary at 100 billion and of Romania at 212 billion.[46] He offered no data on gas reserves in Bulgaria and the GDR. There are several Western estimates of Eastern European oil and gas reserves, based mainly on United Nations and OECD data. The range of estimates made by Western analysts is detailed in table 4.5. The striking features of these analyses is the extent of their disparity, and their divergence from the aforementioned Soviet estimates. Polach rightly pointed to the Comecon practice of evaluating reserves on the basis of geological surveys and production technology existing at the time of the evaluation, a method which ignored the potentially modifying influence of the domestic and export prices of alternative fuels on production levels.[47] Thus what would be defined as recoverable in a Comecon estimate differs from the conclusions that would be reached in evaluating the potential of a Western hydrocarbon province, where, among other factors, likely forward trading prices on the Rotterdam market would be taken into account. Moreover the former system of estimation is subject to revision only in the light of improvements in survey and production techniques, the latter would in addition be influenced by periodic re-evaluation of the forward price. In short, the problem presented by Western analysis is that the basis for evaluation is obscure, with the result that it is not possible to gauge the degree of consistency with, or divergence from, the standard Comecom system of reserve classification.

For the purposes of this study the relevance of the reserve position is the extent to which its perception by Comecon planners influences energy planning

Table 4.5 Western Estimates of Eastern European Oil and Gas Reserves (oil – million tonnes, gas – billion cubic metres)

Country	*A*		*B*		*C*	
	Oil	*Gas*	*Oil*	*Gas*	*Oil*	*Gas*
Bulgaria	7.1	25-30	38	31	5	25
Hungary	21.8	24-50	37	119	20	100
GDR	1.0	5.0	2	14	1	100
Poland	3.6	5-46	8	90	5	130
Romania	158.6	200-246	100	250	165	250
Czechoslovakia	1.9	15.0	2	11	2	15

Sources: (A) J G Polach, 'The Development of Energy in East Europe', in *Economic Development in Countries of Eastern Europe*, Washington DC, US Congress, Joint Economic Committee 1970, pp 376-377; (B) J Bethkenhagen, *Bedeutung und Möglichkeiten des Ost-West-Handels mit Energierohstoffen*, (Deutsches Institut für Wirtschaftsforschung, Sonderheft 104), Berlin, Duncker & Humblot, 1975, pp 85, 88; (C) J R Lee, 'Petroleum Supply Problems in Eastern Europe', in *Reorientation and Commercial Relations of the Economies of Eastern Europe*, Washington DC, US Congress Joint Economic Committee, 1974, p 408

as a whole, and the development of the indigenous oil and gas industry in particular. On balance, it can be said that the available reserves of energy, unevenly distributed in Eastern Europe, allow very little flexibility in planning at a time when the need to increase the contribution of indigenous production to the changing energy balance has become more urgent.

The Eastern European Oil and Gas Industries to 1975

Table 4.6 details the production of major fuels in Eastern Europe from 1971 to 1975. Examination of the trends in hydrocarbon fuels reveals that only in the case of Romanian gas has there been significant growth during this period. The remaining countries show minor gains or losses on generally low levels of production. At the outset of the 1971-1975 Plan decisions were taken that provided for a growing share of oil and gas within the energy balance of each of the member-countries: however, the effect of price rises for oil imported from the Middle East and, from 1 January 1975 from the Soviet Union, have prompted Eastern European planners to look more closely at the possibilities of raising indigenous production of oil and gas.

Bulgaria

Bulgaria is exceptionally poorly endowed with energy resources. According to a 1968 estimate some 82 per cent of energy reserves consist of brown coal, of which two thirds are concentrated in the Marbak-Iztok area: such reserves of hard coal as do exist are said to be insignificant.[48] The trend anticipated in 1970 for the Bulgarian energy balance to 1980 is detailed in table 4.7, and is based on the expectation of increasing imports of oil and gas. It was estimated in 1970

Table 4.6 Eastern European Production of Major Fuels 1971-1975

Country	*Year*	*Oil (thousand tonnes)*	*Natural Gas (million cubic metres)*	*Hard Coal (thousand tonnes)*	*Brown Coal (thousand tonnes)*
Bulgaria	1971	305	327	389	26,620
	1972	249	221	384	26,893
	1973	190	222	351	26,459
	1974	144	180	307	23,998
	1975	122	111	330	27,515
Hungary	1971	1,955	3,705	3,941	23,484
	1972	1,977	4,110	3,671	22,171
	1973	1,989	4,821	3,410	23,371
	1974	1,997	5,101	3,209	22,552
	1975	2,006	5,175	3,021	21,867
GDR	1971	<100	2,853	857	262,814
	1972	<100	5,055	815	248,416
	1973	<100	7,012	753	246,245
	1974	<100	7,732	594	243,468
	1975	100	9,000	539	246,706
Poland	1971	395	5,164	145,491	34,517
	1972	347	5,601	150,697	38,221
	1973	392	5,811	156,630	39,215
	1974	550	5,528	162,202	39,826
	1975	553	5,776	171,625	39,865
Romania	1971	13,793	25,605	6,793	13,808
	1972	14,128	26,552	6,612	16,547
	1973	14,287	28,005	7,172	17,679
	1974	14,486	28,852	7,109	19,789
	1975	14,590	31,570	7,320	19,771
Czechoslovakia	1971	194	1,222	28,818	84,162
	1972	191	1,163	27,925	84,930
	1973	171	1,042	27,779	81,249
	1974	149	976	27,972	82,165
	1975	142	929	28,819	86,272

Sources: *Statisticheskii ezhegodnik stran-chlenov SEV 1976*, pp 77-78; *United Nations Statistical Yearbook 1975*, pp 191, 193 (GDR oil and gas production, not reported in Comecon or own statistical yearbook); J Bethkenhagen, *Bedeutung und Möglichkeiten des Ost-West-Handels mit Energierohstoffen*, (Deutches Institut für Wirtschaftsforschung, Sonderheft 104), Berlin, Duncker & Humblot 1975, pp 87, 90

that demand for oil would rise to 18-20 million tonnes and calculations from data in table 4.8 suggest that gas demand in 1980 would be of the order of 8 billion cubic metres.[49]

The first significant discovery of oil in Bulgaria was that of the Tyulen field in 1954. Its reserves, estimated at 3.5 million tonnes, were exploited rapidly and the field reached peak production in 1957, since when it has shown a gradual decline.

In the early sixties the Gigen and Dolnyi-Dybnik fields were discovered, the

Table 4.7 Bulgarian Energy Balance, 1960-1980, Estimated in 1970 (%)

Fuel	*1960*	*1970*	*1980*
Solid Fuel, All Types	81.6	53.1	29.6
Oil	16.0	43.3	49.0
Gas	–	2.5	16.4
Other (Hydro & Nuclear)	2.4	1.1	5.0

Source: T Khristov, 'Novi tendentsii v razvitieto na energetikata v Bolgariya', *Geografiya (Sofia)*, 8/1970, p 2

former yielding a heavy, viscous crude oil, the latter a light type. By the late sixties the Gornyi-Dybnik field had been added, and a single gas condensate field, the Chiren, was discovered in 1963.[50] Oil refining became concentrated in two complexes, at Pleven and Burgas. The crude oil distillation capacity at Burgas in 1971 was 7 million tonnes per annum, and this was scheduled to rise to 12-13 million by 1975, with the prospect of further expansion to 18-20 million by 1980.[51] The relative scarcity of domestic reserves of oil prompted Bulgarian planners to consider critically their use of refined products. In particular, evaluation of the role of fuel oil in relation to coal illustrates the planners' appreciation of the importance of maximizing the output of light products. Until the early sixties practically all of Bulgaria's gasoline and kerosene requirement were imported, mainly from the USSR and to a lesser extent from Romania. The bulk of domestic refinery output consisted of middle distillates and fuel oil. By the late sixties it was decided to raise the production of gasoline within total refinery output from zero (1958) to 20.5 per cent by 1975, and middle distillates from 22.5 to 28.2 per cent in the same period, counterbalanced by a fall in the fuel oil share from 51.8 to 45.8 per cent. Balancing imports of refined products would evidence a relative decline in two former product groups and an increase in the latter, but within a smaller total product import requirement.[52]

The discovery of five gasfields in the early fifties during exploration for oil gave some cause for optimism about further development in Bulgaria. However, the Chiren gas condensate deposit was not brought into production until 1965. This gas was found to have a particularly high methane content (92 per cent), little impurity and an associated yield of 50 cubic centimetres of low octane condensate per thousand cubic metres' gas production, usable either in the manufacture of gasoline or as petrochemical feedstock.[53] On the basis of the availability of methane and condensate from the Chiren deposit, it was decided to construct a petrochemical complex at Vratsa. In 1969 the Devetash gasfield was discovered. This gas contained 60-80 per cent methane, with an associated yield of 150 cubic centimetres of condensate per thousand cubic metres of gas. As a result of this discovery it was estimated that reserves in the categories A+B+C1 could be expected to rise to between 10 and 15 billion cubic metres by 1975, and that a production level of 1 billion cubic metres could be sustained. This would result in an increasing share of indigenous gas in the energy balance. In 1970 it was thought likely that oil and gas, which in 1968 accounted for 42 per cent of the Bulgarian energy balance, would rise to 60 per cent in 1975, with the prospect of attaining 65-70 per cent in 1980.[54] However, the Bulgarian gas industry experienced rapid resource depletion during the 1971-1975 period which was reflected in declining production. The long-term position of gas in the energy balance is acknowledged to depend on supply from the Soviet Union, facilitated by Bulgarian provision of equipment and manpower in joint projects.

The effect of the increased availability of gas has been that of hastening the move away from fuel oil and complementing the use of oil in energy-intensive industries such as metallurgy and power generation.[55] During the 1971-1975 Bulgarian Plan indigenous coal was directed for the most part into the fuelling of power stations, this sector accounted for approximately 75 per cent of coal consumption. At the end of the Plan reserves were still considered adequate to increase coal utilization in this outlet.[56] In contrast the discovery of new reserves of oil and gas in this period was well below the increase in their consumption, and in view of this declining production increased import dependency was acknowledged as inevitable.[57] The balance between indigenous production of oil and gas and the quantities supplied for consumption from 1971 to 1975 is detailed in table 4.8. It can be seen that the total available for 1975 was below that for 1974, evidence of the conservation measures introduced after the doubling of oil prices by the Soviet Union in February 1975, and reflecting a generally pessimistic outlook for the oil and gas industries.

Table 4.8 Availability of Oil and Natural Gas in Bulgaria 1971-1975

A Oil (thousand tonnes)

	1971	*1972*	*1973*	*1974*	*1975*
Indigenous Production	305	249	190	144	122
Imports	7,547	8,279	9,652	10,629	10,459
Availability	7,842	8,518	9,742	10,763	10,581
Import Dependency (%)	96.2	97.2	99.1	98.8	98.8

B Gas (million cubic metres)

	1971	*1972*	*1973*	*1974*	*1975*
Indigenous Production	327	221	222	180	111
Imports	–	–	–	307	1,185
Availability	327	221	222	487	1,296
Import Dependency (%)	–	–	–	63.0	91.4

Sources: *Statisticheskii ezhegodnik stran-chlenov SEV 1976*, pp 78, 354; *Vneshnyaya torgovlya SSSR 1975*, p 121

Hungary

It is revealing to compare two estimates of the development of the Hungarian energy balance, one written in July 1970, the other in October 1976. Despite substantial increases in the price of imported oil that took place between these estimates, and the logistic difficulties faced by the Soviet Union (Hungary's principal oil supplier), the estimates assign only slightly differing shares to each fuel, as shown in table 4.9.

Table 4.9 Analyses of Hungarian Energy Balance 1965-1980 (%)

Fuel	*1965A*	*1970A*	*1970B*	*1975A*	*1975B*	*1980A*	*1980B*
Coal	66.7	48.9	50.0	37.9	36.0	27.1	27.0
Oil	21.4	29.9		32.4		39.4	
			43.0		57.0		65.0
Gas	6.1	13.3		21.6		22.2	
Other	5.8	7.9	7.0	8.1	7.0	11.3	8.0

Sources: (A) J Kalman Joó, 'Wärmeversorgung von Städten in der Ungarischen Volksrepublik', *Energietechnik*, 7/1970, p 322; (B) V Zurbuchen, 'Die 5 Fünfjahrplan der Ungarischen Volksrepublik 1976 bis 1980', *Wirtschaftswissenschaft*, 10/1976, p 1325

One of the influential factors is that while Hungary has no substantial reserves of any one fuel, it has limited reserves of each, enabling planners to enjoy some measure of short-term flexibility. Lisichkin for example, estimated that Hungary would have 55 per cent self-sufficiency in energy in 1975 and not less than 50 per cent in 1980.[58]

In the sixties there were a number of oil and gas discoveries in Hungary, including the relatively large Aldyo field. The annual production of 1 million tonnes of oil and 2 billion cubic metres of gas from this field was felt to be sustainable for 20 to 30 years. In 1974 there were a total of 34 producing oilfields in Hungary.[59]

Of the five refineries in operation in 1970 the most complex was the integrated petroleum and petrochemical plant at Szazhalombatta, the capacity of which was planned to reach 6 million tonnes per annum in 1975. The other major refinery is situated at Leninvaros, where capacity of 6 million tonnes per year is planned for 1980. Total Hungarian capacity is scheduled to rise from 10 million tonnes per annum in 1975 to 16 million in 1980.[60]

The gas industry in Hungary showed a substantial degree of growth during the 1971-1975 Plan, and recorded an increase of 49 per cent in 1975 over the 1970 level. The impact of the Soviet Union as a gas supplier will be felt after 1978, when the major pipeline from the Orenburg gasfield is extended into Hungary. Availability of oil and gas in Hungary in the period from 1971 to 1975 is detailed in table 4.10.

Despite the continued direction of capital and human resources into exploratory work in the oil and gas sectors, the prevailing view at the outset of the 1971-1975 Plan was that Hungarian oil production would peak at around 2 million tonnes per annum and gas production at a little over 6 billion cubic metres. Without further discoveries of the size of the Aldyo field, it was considered that production of both oil and gas was likely to start a decline by 1980. At the same time forward estimates for 1975 demand, indicated a requirement of 10 million tonnes of oil and 7 billion cubic metres of gas, with the expectation that the rising trend would continue.[61]

A significant feature of the Hungarian oil and gas industries is that despite capital investment in oil and gas development of some 23 billion forints, 60 per cent higher in 1971-1975 than in the previous five year period, the decline in oil

Table 4.10 Availability of Oil and Natural Gas in Hungary 1971-1975

A. Oil (thousand tonnes)

	1971	*1972*	*1973*	*1974*	*1975*
Indigenous Production	1,955	1,977	1,989	1,997	2,006
Imports	4,892	6,065	6,555	6,817	8,432
Availability	6,847	8,042	8,544	8,814	10,438
Import Dependency (%)	71.4	75.4	76.7	77.3	80.8

B. Gas (million cubic metres)

	1971	*1972*	*1973*	*1974*	*1975*
Indigenous Production	3,705	4,110	4,821	5,101	5,175
Imports	208	200	200	200	806
Availability	3,913	4,310	5,021	5,301	5,981
Import Dependency (%)	5.3	4.6	4.0	3.8	13.5

Source: *Statisticheskii ezhegodnik stran-chlenov SEV 1976*, pp 78, 363

production was not averted but gas production rose rapidly. The gas industry has therefore come to be regarded as the most promising of the fuels industries for future development.

The inevitability of increasing exports of oil not only from the Soviet Union but also from OPEC producers prompted Hungarian collaboration with Yugoslavia and Czechoslovakia in the 'Adria' pipeline project (see Appendix A) to facilitate the supply of oil from Kuwait. The agreement reached before the price rises of 1973-1974, envisaged OPEC oil eventually accounting for 10-12 per cent of Hungary's oil demand.[63] This high level is now thought unlikely, and the Soviet Union is viewed as the preferred supplier.[64] Towards the end of the 1971-1975 Plan it was decided to attempt the arrest of the declining trend in domestic production of coal[65] to compensate for the likelihood of static oil production, at approximately 2 million tonnes per annum, during the 1976-1980 period. Half this amount would be provided by a single field, the Aldyo.[66]

As with the Soviet Union, the relatively late development of the gas industry and prospects for its further development have meant that there was some scope for alleviating the deteriorating balance of supply and demand for oil. In 1970 it was estimated that demand for gas would rise to 8-9 billion cubic metres by 1980.[67] Bearing in mind the increasing domestic production and greater delivery capacity for Soviet gas which is scheduled to become operational in 1978, this figure could well be exceeded. Whereas during the sixties gas was directed to the petrochemical sector and to selected branches of industry, during the 1971-1975 period its use as a domestic fuel increased and this trend is expected to continue.[68] According to a recent Hungarian source the emergence of gas within the energy balance from 1971 to 1975 has had the effect of substituting for the import of some 600 thousand tonnes of oil.[69]

The GDR

The energy sector of the GDR economy has been and remains predominantly based on the production and utilization of brown coal. Ultimate reserves of this fuel are put at 49 billion tonnes, of which 25 billion are amenable to open-cast mining. Some 60 per cent of brown coal reserves are concentrated in deposits the individual potential of which runs to over 200 million tonnes. The basic disadvantages of brown coal are its low calorific value by weight (on average 2,000 to 2,800 kilocalories per tonne, or approximately half that of hard coal), and high moisture content, ranging from 45 to 65 per cent. Reserves of hard coal in the GDR total approximately 50 million tonnes, and these are concentrated in a single deposit at Zwickau. Production of hard coal in the GDR has made a small and declining contribution to the energy balance in the seventies.[70]

Exploratory work for oil and gas took place during the sixties with limited success in the gas sector. Discoveries were made in the Magdeburg area but this find proved disappointing. The gas showed only a 30 per cent methane content,[71] lowering the calorific value of the gas and necessitating investment in further processing facilities.

The 1971-1975 Plan for the GDR, approved and adopted in December 1971, provided for the allocation of some 30 per cent of total industrial investment to the fuels and energy industries. Output of brown coal was scheduled to rise to 255 million tonnes per year by 1975 and oil refining capacity from 10.6 million tonnes per year in 1970 to 18 million in 1975. It was also anticipated that by 1975 deliveries of Soviet gas would have commenced.[72] GDR planners wished to bring about the greater use of oil and gas for reasons of energy efficiency, and as the basis of the further development of the petrochemical industry. They also anticipated growing problems in the indigenous coal industry, not least of which were the ratio of over-burden to recovered coal, estimated as likely to increase from 3.5:1 at the end of 1971 to over 5:1 by 1980, coupled with the consequent high level of demand for water-cleansing facilities.

During the 1971-1975 period oil production in the GDR remained at a very low level in relation to imports and consumption. The bulk of supply was provided by the Soviet Union, supplemented by small quantities from Middle East producers (see table 4.11). During the sixties the hydrogenation of brown coal to produce petroleum products yielded about 1 million tonnes per year but this process was abandoned as uneconomic by 1970.[73] The sole producing oilfield in the GDR is the Reinkenhagen, near the Baltic coast, and the geology of this field has prompted the commencement of offshore exploratory drilling.[74] However, indigenous production of oil is not expected to make a significant contribution to energy demand in the GDR in either the short- or long-term.

In contrast the emergent, though limited contribution and potential of the indigenous gas industry, supplemented increasingly by imports from the Soviet Union, have generated the major change in the development of the GDR energy balance from 1971 to 1975, and it is likely that this contribution will grow (see (see table 4.11). The most awkward problem faced in the utilization of indigenous gas is that of separating the substantial quantities of nitrogen found in association with methane. However, despite this there are still savings to be made from

Table 4.11 Availability of Oil and Natural Gas in the GDR 1971-1975

A. Oil (thousand tonnes)

	1971	*1972*	*1973*	*1974*	*1975*
Indigenous Production	neg	neg	neg	neg	100
Imports	10,919	14,858	16,045	16,434	16,997
Availability	10,919	14,858	16,045	16,434	17,097
Import Dependency (%)	100	100	100	100	99.5

neg = negligible

B. Gas (million cubic metres)

	1971	*1972*	*1973*	*1974*	*1975*
Indigenous Production	2,853	5,055	7,012	7,732	9,000
Imports	110	5	760	2,841	3,226
Availability	2,963	5,060	7,772	10,573	12,226
Import Dependency (%)	3.7	0.1	9.8	26.9	26.4

Sources: *Table 4.6*; *Statisticheskii ezhegodnik stran-chlenov SEV 1976*, p 371; V P Maksakovskii, *Toplivnaya promyshlennost' sotsialisticheskikh stran Evropy*, Moscow, Nedra, 1975, p 187

the use of gas, since in certain processes, for example metal processing and glass manufacture, it is up to 25 per cent more efficient than brown coal. There are usually additional savings in that gas-fuelled processes are generally the least labour-intensive.[75]

Towards the end of the 1971-1975 period an East German analyst pointed to two major considerations affecting the development of the energy balance. The increasing need to reduce dependence on indigenous brown coal to the major energy source, by expanding the use of natural gas, and the increasing dependence on imports in order to achieve greater efficiency in energy utilization.[76]

An analysis of the GDR energy balance from 1960 to 1975 is given in table 4.12, showing the increasing share of oil and gas. Between 1970 and 1975 the increase in percentage share was much the same for both oil and gas; the share of gas showed the major increase between 1974 and 1975, when deliveries commenced from the Soviet Union. The poor hydrocarbon resource base has necessitated the use of oil and gas only in those processes in which there is a

Table 4.12 The GDR Energy Balance, 1960-1975 (%)

Fuel Type	*1960*	*1965*	*1970*	*1975*
Solid Fuel	96.8	93.0	85.7	70.9
(of which, brown coal)	(87.5)	(84.2)	(78.2)	(65.0)
Oil	2.5	6.4	13.1	20.1
Gas	–	–	0.7	7.5
Nuclear	–	–	0.2	1.3
Other	0.7	0.6	0.3	0.2

Source: H Wambutt, 'Planmässige Entwicklung der Energiewirtschaft der DDR', *Einheit*, 6/1974, p 706

substantial advantage over coal. Thus, for example, some 83 per cent of electricity generated in 1973 was fired on brown coal,[77] and it is intended to continue the use of brown coal in such basic processes as power generation and general industrial steam-raising. However, there are limits to the applicability of this policy in the medium-term on environmental and economic grounds. The majority of brown coal reserves are concentrated in the Cottbus district (58 per cent), around Leipzig (25 per cent) and Halle (12 per cent), and the impact on the environment arising from further development of open-cast mining, together with the expectation of rising production costs are gaining increasing influence amongst policymakers.[78]

It is nonetheless intended to arrest the decline in the production of brown coal and to attempt the maintenance of production at least at the 1975 level during the 1976-1980 Plan.[79] The contribution of nuclear power to the GDR energy balance during 1971-1975 was minimal: no substantial increase is expected by 1980 on the basis of developments carried out by 1975.

Poland

As a result of its extensive reserves of hard and brown coal and a highly developed mining industry Poland has a net energy surplus, and is an established trader in coal both in Comecon and in the world market. Total explored reserves of hard coal have been put at 200 billion tonnes, of which categories A+B+C1 account for some 80 billion.[80] The dominant role of coal in the Polish energy balance in the period 1961 to 1975 is detailed in table 4.13. Throughout the sixties consumption of oil and gas increased as a result of the exploitation of indigenous reserves in the Central Carpathian basin, but more significantly on account of imports from the Soviet Union. Production of oil in Poland grew from 1971 to 1975, but it still makes a relatively small contribution to total energy supply. The dispersion of hydrocarbon resources in Poland is given in table 4.14.

Exploratory work carried out during the 1971-1975 Plan was centered on the Carpathian basin with greater success in the gas rather than in the oil sector. This inevitably prompted planners to favour a conservationist policy in depletion of domestic oil reserves.[81] However, the OPEC price rises, compounded by the Soviet price rise, required that this policy be re-assessed. As far as the internal

Table 4.13 The Polish Energy Balance 1961-1975 (%)

Fuel	*1961*	*1965*	*1970*	*1975*
Coal	90.7	88.2	81.4	77.9
Oil	4.4	6.4	9.7	14.9
Natural Gas	1.4	2.3	6.3	6.7
Peat and Wood	3.4	3.0	2.0	
Hydroelectricity	0.1	0.1	0.6	0.5
Nuclear	–	–	–	

Sources: (i) S M Lisichkin, *Energeticheskie resursy i neftegazovaya promyshlennost' mira*, Moscow, Nedra, 1974, p 88;
(ii) J Bethkenhagen, 'Die Zusammenarbeit der RGW-Länder auf dem Energiesektor', *Osteuropa Wirtschaft*, 2/1977, p 77

Table 4.14 Distribution of Polish Hydrocarbon Reserves in Categories A+B+C1 (%)

	Oil	*Natural Gas*
Polish Lowlands	13.5	32.9
Precarpathian	38.7	64.9
Carpathian	47.8	2.2

Source: J Pilch, 'Stan i perspektywy rozwoju górnictwa naftowego', *Wiadomości naftowe*, 3/1973, p 55

logistic system permitted, it proved possible to raise domestic production during the latter half of the 1971-1975 Plan. The Poles faced a similar problem to the Soviet Union in seeking to raise oil production, namely that of underperformance in the discovery of new reserves,[82] plus the high costs of exploiting domestic oil, which on a standard fuel basis proved to be ten times that of natural gas (see table 4.15).[83]

The growth of natural gas within the Polish energy balance, both as a result of the development of indigenous reserves and of imports from the Soviet Union, has been slight in comparison with other Eastern European countries. At the outset of the 1971-1975 Plan it was estimated that the share of natural gas in the energy balance would rise from 6 per cent in 1970 to 9 per cent in 1975,[84] out of a total energy demand rising from 132.8 million tonnes of standard fuel in 1970 to 146 million in 1975.[85] The geographical distribution of planned production is given in table 4.16. A major influence on the underfulfilment of the oil production plan was that deeper drilling proved necessary in order to discover new fields and to bring existing deposits into production. This resulted in extensive delays. In the gas sector, certain of the additional discoveries were found to contain gas with a lower methane content than originally anticipated.[86] Gas from the Sudetan lowlands was found to have a particularly high nitrogen

Table 4.15 Availability of Oil and Natural Gas in Poland 1971-1975

A. Oil (thousand tonnes)

	1971	*1972*	*1973*	*1974*	*1975*
Indigenous Production	395	347	392	550	553
Imports	7,894	9,703	11,140	10,582	13,306
Availability	8,289	10,050	11,532	11,132	13,859
Import Dependency (%)	95.2	96.5	96.6	95.1	96.0

B. Gas (million cubic metres)

	1971	*1972*	*1973*	*1974*	*1975*
Indigenous Production	5,164	5,601	5,811	5,528	5,776
Imports	1,493	1,505	1,715	2,123	2,516
Availability	6,657	7,106	7,526	7,651	8,292
Import Dependency (%)	22.4	21.2	22.8	27.7	30.3

Source: *Statisticheskii ezhegodnik stran-chlenov SEV 1976*, pp 78, 386

Table 4.16 Geographical Distribution of Planned Gas Production in Poland 1971-1975 (billion cubic metres)

Region	*1971*	*1972*	*1973*	*1974*	*1975*
Carpathians and Carpathian Foothills	4.9	5.0	5.14	5.14	5.2
Sub-Sudeten Monocline	0.1	0.6	2.10	3.38	5.2
Other Lowlands	–	–	–	0.20	0.6
Total	5.0	5.6	7.24	8.72	11.0

Source: *Gas World*, July 1974, p 384

content, and the correspondingly lower level of methane extraction accounted for a recorded shortfall against Plan for gas production.[87] Only the desirability of, and facilities for extraction of helium, also present in Sudeten gas, rendered exploitation of this field economic. In view of the fact that availability of gas was lower than planned between 1971 and 1975, preferential allocation was given to the petrochemical industry, to which it was estimated that, in 1975,[88] some 3.74 billion cubic metres would be directed.

However, in the aftermath of the oil price rises there was a resurgence of interest in the development of indigenous resources and the area considered most likely to yield new oil and gas deposits was the Baltic continental shelf. On 24 November 1975 a joint prospecting agreement was signed between Poland, the GDR and the Soviet Union, under the terms of which the organization, Petrobaltic, with headquarters in Gdansk, would coordinate exploration in the Baltic Sea.[89]

Refinery capacity in Poland has shown an increase from 9 million tonnes per year in 1970 to 13.5 million in 1975. This met Poland's need for refined products.[90] At the end of 1975 there were nine refineries in Poland, the largest at Plock fed with Soviet oil, and one at Gdansk constructed specifically to process Middle East oil. The latter had been running at below design capacity since it was commissioned prior to the price rises of 1973/1974, when projections of future imports of oil were based on considerably lower buying prices. During 1971-1975 the Polish refinery balance showed a predominance of fuel oil, though during this period its share declined. Given the tightening balance between supply and demand for oil, increasing motorization within the economy and the growing need for oil-derived feedstock for the petrochemical industry, the trend towards maximizing the production of gasoline, naphtha and middle distillates that commenced in the early seventies was expected to intensify.[91]

Between 1971 and 1975 Poland did not face an energy shortage: however, there have been difficulties in the domestic oil and gas industries and these were compounded by the Soviet Union's inability to increase the supply of oil and gas above contractual commitments, due to its own technical and logistic problems. Poland faced particular difficulties at the time of the 1973 'oil crisis' and subsequently, in that it was unable to compensate for immediate shortcomings by substantially increasing imports from OPEC producers. Hence in the latter part of the Plan Poland was obliged to introduce measures for fuel, and especially for oil, conservation.

Romania

Romania is particularly well provided with energy resources. Total coal reserves are put at 6 billion tonnes, of which 3.7 billion consist of brown coal. Approximately 90 per cent of these reserves are concentrated in the South-West of Romania in Oltenia. This resource concentration simplifies transportation, but poses increasingly difficult problems of environmental deterioration. In 1970, open-cast mining provided 33.8 per cent of coal production. Brown coal is used predominantly in electricity generation, and in 1974 it was estimated that some 70 per cent of the production during 1971-1975 would be directed to this use.[92] Moreover, in contrast with the rest of Eastern Europe the Romanian energy balance was based largely on oil and gas from an early stage. From 1950 to 1965 indigenous oil and gas together accounted for over 70 per cent of Romanian energy consumption.[93] In the period from 1950 to 1970 investment in the oil and gas industries totalled 64 billion lei, some 19 per cent of overall industrial investment.[94] Total oil reserves are now put at 100-130 million tonnes,[95] reserves of gas in the categories A+B+C1 at 212 billion cubic metres.[96]

During the sixties indigenous production of oil exceeded domestic demand, and Romania was able to develop export trade in refined petroleum products. However, in the 1971-1975 period production of oil did not increase significantly whereas demand for refined products did. Moreover on the basis of the expanding exports of refined products and in anticipation of the pre-crisis price for OPEC oil of 2-3 dollars per barrel being maintained in real terms, thus allowing Romania to export refined products to an expanding market, it was decided in 1975 to expand Romanian refining capacity to 25 million tonnes per year.[97] Extraction of natural and wellhead gas in Romania continued to rise, recording a 31.6 per cent increase between 1970 and 1975. The value of Romanian natural gas as a fuel is enhanced by its exceptionally high methane content, at 98 to 99.5 per cent, the highest in Comecon.[98]

Energy consumption in Romania rose from 22.8 million tonnes of standard fuel in 1958 to 38.7 in 1965, and it was expected to reach 78-80 million in 1980.[99] The relative importance of individual fuels in the energy balance from 1950 to 1970 is shown in table 4.17: oil's share declined, which is atypical of Comecon, and greater significance was given to gas. This was as a result of a deteriorating ratio of oil reserves to production and an inability to expand imports from OPEC beyond the counterbalancing of exports of petroleum products. The pattern of supply of crude oil in Romania is detailed in table 4.18.

The major part of the Romanian development effort has been rapidly directed towards the application of improved technology, which can increase economically the amount of oil recovered from a single field. By the end of 1975 the national average level of recovery had reached 30 per cent of oil in place, the resulting production level just covering domestic need. However, in view of depletion and a probable decline in the number of deposits being worked it was estimated that a recovery level of 42 per cent must be attained by 1980.[100]

For the most part Romanian oil is of the light, paraffinic variety, suitable for maximum output of gasoline, naphtha and middle distillates.[101] The preference of planners for a slowdown in the rate of growth in consumption of oil as fuel,

Table 4.17 The Romanian Energy Balance Balance 1950-1970 (%)

Fuel	*1950*	*1955*	*1960*	*1963*	*1970*
Coal	14.9	11.3	12.1	12.7	16.7
Oil	47.8	54.0	48.7	41.6	20.5
Gas	26.4	26.9	34.0	41.0	55.6
Wood/Peat	9.5	6.4	4.1	3.1	5.5
Hydroelectricity	1.4	1.4	1.1	1.6	1.7

Sources: F W Carter, 'Natural Gas in Romania', *Geography*, April 1970, p 220; I V Herescu, 'Creşterea economica şi consumul de energie', *Revista economică*, 34/1975, p 13

Table 4.18 Availability of Oil in Romania 1971-1975 (thousand tonnes)

	1971	*1972*	*1973*	*1974*	*1975*
Indigenous Production	13,793	14,128	14,287	14,486	14,590
Imports	2,858	2,873	4,143	4,538	5,085
Availability	15,651	17,101	18,330	18,924	19,675
Import Dependency (%)	18.3	16.8	22.6	24.0	25.8

Source: *Statisticheskii ezhegodnik stran-chlenov SEV 1976*, pp 78, 391

in order to facilitate its increased use as petrochemical feedstock influenced the original targets set for production to 1975. It was intended that production in 1975 should be 13.1 to 13.5 million tonnes, or 98.5 to 101.5 per cent of the 1970 level, with the rising demand for fuel being met by gas and coal. However, this target was raised prior to the ratification of the 1971-1975 Plan, which then set the 1975 target at 14.5 million tonnes, a figure which was exceeded.[102] Discussions on energy policy prior to the formulation of the Plan for 1976-1980 indicated a similar preference for a low growth rate in oil production. In 1974, the optimal level for oil production in 1980 was put at 15.5 million tonnes.[103]

Changes in the pattern of oil refining date from the early sixties, when the Romanian petrochemical industry began to grow and the ratio of proven reserves to production commenced its decline. From 1963 the refinery at Brazi underwent expansion, with added cracking and hydrofining capacity and a powerforming unit providing greater output of high octane gasolines and naphtha. The refinery at Ploieşti was also expanded and modified so that output was much the same as that of the Brazi complex. A new refinery was built at Piteşti, designed to process imported oil, in which sulphur levels were higher than in the indigenous varieties.[104] The resulting changes in output of refined products, in particular the substantial growth of middle distillate production, is detailed in table 4.19.

The gas industry continued to grow and the demands of the chemical and petrochemical industries were an instrumental factor in prompting its further rapid development. At the tenth Congress of the Romanian Communist Party, held in 1969, a level of output for the chemical industry in 1975 was foreseen as some 82-92 per cent above the 1965 level, and the production level of natural gas, commensurate with this expansion and with growth in demand for energy

Table 4.19 Changes in the Romanian Refinery Balance 1950-1970 (thousand tonnes)

	1950	*1960*	*1970*
Gasoline	1,502	2,792	2,786
Middle Distillates	731	2,376	4,979
Fuel Oil	1,681	3,824	4,249
Lubricants	125	311	546
Bitumen	92	249	537
LPG	12	77	207
Aromatics	–	0.5	358

Source: G Pacoste, 'Producția și utilizarea țițeiului și a gazelor naturale în România', *Energetica (Bucharest)*, 6/1971, p 259

generally, in the range of 22-24 billion cubic metres, supplemented by some 5 billion cubic metres of wellhead gas.[105] In fact domestic production of natural and wellhead gas showed a growth of 7.6 billion cubic metres between 1970 and 1975, rising from 24 billion to 31.6 billion.[106] There were no imports.

Though Romania has been able to pursue a policy of energy autarchy up to 1975, the short-term position is one of tightening supply, especially in the oil industry. Thus when OPEC increased the price of oil in 1973 and 1974, Romania's decision to impose conservation measures was prompted not by the threat of an energy shortage, nor for reasons of an immediate currency drain in maintaining imports (the cost of these was largely recouped by the export of refined products), but by the need to conserve domestic resources. These conservation measures were imposed while the long-term impact of higher Middle East oil price on future energy and chemical trade policy was evaluated.[107] Planners are conscious of the fact that from 1960 to 1967 hydrocarbon discoveries were believed to constitute some 86 per cent of known reserves[108] and the consequent rehabilitation of coal has become a feature of Romanian energy planning. Domestic reserves of coal are considered sufficient to sustain a higher production level than is presently the case, and there are possibilities of joint developments in coal production, both within and outside Comecon. These would go some way towards alleviating the difficulties that have arisen in the oil industry.[109] The development of the Romanian energy balance between 1970 and 1975 is outlined in table 4.20.

Table 4.20 The Romanian Energy Balance 1970-1975 (%)

Fuel	*1970*	*1973*	*1975*
Coal	16.7	20.1	23.5
Oil	20.5	27.4	34.3
Gas	55.6	45.9	36.3
Other	7.2	6.6	5.9

Source: *Table 4.17*; I V Herescu, 'Dezvoltarea bazei energetice(l)', *Revista economică*, 28/1976, p 1

Czechoslovakia

The problems Czechoslovakia faces in relation to energy supply are severe. Reserves of each conventional fuel are limited: the basic one is coal, total reserves of which are put at 13.7 billion tonnes at depths up to 1,800 metres, this includes 6.5 billion below 1,200 metres. Of the 13.7 billion tonnes brown coal accounts for 8.2 billion, and proven reserves are considered sufficient to sustain anticipated production for several decades.[110] Reserves of oil and gas are said to be insignificant.[111] The development of the Czech energy balance between 1960 and 1970 (detailed in table 4.21) shows that total consumption of energy rose from 61.9 million tonnes of standard fuel to 90.6 million during this period.[112]

Table 4.21 The Czechoslovak Energy Balance 1960-1970 (%)

Fuel	*1960*	*1965*	*1970*
Solid Fuel	89.1	83.6	74.9
Oil	6.2	11.5	17.8
Gas	2.6	1.4	3.3
Other	2.1	3.5	4.0

Source: K Houdek, 'Palivoenergetická základna CSSR v páté petiletce', *Plánované hospodárství*, 1/1972, p 12

Oil production in Czechoslovakia consists of the combined output of a number of small fields. Although on a standard fuel basis oil is the highest cost fuel produced, its conversion efficiency in many industrial processes justifies its production.[113] Following the opening of the *Druzhba* pipeline in 1964, Czech imports of Soviet oil grew rapidly, and throughout the 1971-1975 Plan Czechoslovakia was the Soviet Union's largest purchaser of oil. Small quantities were purchased from OPEC producers and during the 1971-1975 period Czechoslovakia agreed to cooperate with Hungary and Yugoslavia in the development of the 'Adria' pipeline (see Appendix A). The availability of oil in Czechoslovakia is detailed in table 4.22. There appears to be no prospect of increasing domestic production of oil in the short term: even current production is likely to continue its decline. Refinery capacity in 1975 was put at 17 million tonnes per year, with refineries at Bratislava and Zaluzi accounting for the bulk of this amount. Gasoline and naphtha have now become the priority products.[114]

Production of natural gas in Czechoslovakia has fluctuated around an annual level of 1 billion cubic metres since the early sixties, and now exhibits a declining trend. Imports from the Soviet Union have increased during the 1971-1975 Plan and these will rise substantially after 1978, when the pipeline from the Orenburg field is scheduled for completion. The supply of gas from 1971 to 1975 is also detailed in table 4.22. As in the case of the oil industry, it is difficult to see how any increase in indigenous production can be expected in the near future. During the 1971-1975 Plan gas had the lowest production cost of any domestic fuel: unfortunately the poor ratio of reserves to production prevents further expansion.[115]

Table 4.22 Availability of Oil and Natural Gas in Czechoslovakia 1971-1975

A. Oil (thousand tonnes)

	1971	*1972*	*1973*	*1974*	*1975*
Indigenous Production	194	191	171	149	142
Imports	11,505	12,571	14,176	14,665	15,839
Availability	11,699	12,762	14,347	14,814	15,981
Import Dependency (%)	98.3	98.5	98.8	99.0	99.1

B. Gas (million cubic metres)

	1971	*1972*	*1973*	*1974*	*1975*
Indigenous Production	1,222	1,163	1,042	976	929
Imports	1,660	1,957	2,385	3,256	3,821
Availability	2,882	3,120	3,427	4,232	4,750
Import Dependency (%)	57.6	62.7	69.6	83.3	80.4

Source: *Statisticheskii ezhegodnik stran-chlenov SEV 1976*, p 78, 405

In a recent work published in Canada the Czech émigré economist B Korda pointed to the economic mechanism itself as the root cause of Czechoslovakia's energy problems.[116] Korda drew attention to the fact that the level of primary energy consumption is high in relation to final consumption, in effect that the economy exhibited low conversion efficiency in the major energy-intensive processes.[117] He attributed this to excessive past reliance on domestic coal, a policy which in his view was motivated by Stalin's desire to decrease Czechoslovakia's trade with the capitalist world. This eventually precluded imports of relatively cheap West German hard coal.[118] He argued that since neither the Soviet Union nor the rest of Comecon could replace the West as a trading partner, Czechoslovakia had no option but to develop domestic coal regardless of cost, given the limitations on oil and gas supply at that time from the Soviet Union. Even when the Soviet export surplus returned in the late fifties, he maintained that contract quantities were insufficient to meet potential demand, and hence Czechoslovakia retained a coal-based energy economy.

The price rises imposed by OPEC and the Soviet Union had a particularly strong impact on Czechoslovakia, and the need to cut plans for the import of oil from the former was acknowledged by a Czech energy planner in 1974.[119] The imposition of energy conservation measures had an immediate though limited effect in improving conversion efficiency through the elimination of the more obvious wasteful practices. However, the main effect of the price rises, particularly of those imposed by OPEC, was that of engendering greater interest on the Czech part in the joint development of Soviet resources. Nuclear power was the particular focus of this interest which Korda saw as the best and most likely solution to Czech energy problems in the long-term.[120] However, the short-term problem is particularly severe, and in order to meet rising energy demand there is no alternative to the further development of the domestic coal industry, despite increasing production costs and the implications for the environment entailed in open-cast mining.[121]

The development of the Czech energy balance from 1970 to 1980 is outlined in table 4.23. The effect of the price rises is shown in the difference between the share of oil anticipated in 1972 and the final share. However, the percentages understate the absolute decline since the total energy consumption figure recorded was lower than envisaged. The low share of gas reflects the difficulties experienced in domestic production and the inability of the Soviet Union to deliver in excess of contracted quantities due to her own problems in gas production.

Table 4.23 The Czechoslovak Energy Balance 1970-1975 (%)

Fuel	*1970*	*1973*	*1975 Anticipated*	*1975 Actual*
Coal	74.9	68.3	63.7	65.9
Oil	17.8	23.9	26.8	24.6
Gas	3.3	4.6	5.4	5.9
Other	4.0	3.2	4.1	3.6

Sources: (1970 and 1975 Anticipated) K Houdek, *op cit*, p 12;
(1973 and 1975 Actual) J Bethkenhagen, *Die Zusammenarbeit der RGW-Länder . . .*, (1977), p 77

Oil and Gas in Eastern Europe to 1975: An Overview

Eastern Europe is facing considerable difficulty in gearing supply and demand for oil and gas within the energy balance, though the size of the difficulty varies in each country. Also the opportunities for solving the problem vary. Between 1971 and 1975 little could be done to counter the effect of the OPEC and Soviet oil price rises and the situation was aggravated by the Soviet Union's inability to raise production and exports substantially to Eastern Europe after the OPEC rises. Korda and Moravcik[122] have suggested that a policy of modernizing productive capacity, so that industrial output might be increased by improving the efficiency of existing facilities rather than by the installation of further energy-intensive processes, may be the best option to counteract tightening supply and rising prices. However, if such a policy were adopted, it would have little effect on the short-term position. Whereas Eastern Europe has sustained a serious blow from the price rises of October 1973 to February 1975, on balance the economic impact has been measurably less serious than on industrialized countries elsewhere. In undertaking a greater commitment of resources to the development of Soviet energy reserves, Eastern Europe does at least have the opportunity of securing a high level of guaranteed supply in the medium-term.

Credits from the Soviet Union will help offset the effect of the oil price rise. Though this puts the Eastern European countries in a more dependent position in relation to the Soviet Union, the financial impact is less harsh than that of raising finance for imports from OPEC. Writing in 1974 the American analyst J R Lee indicated that the import of 50 million tonnes of OPEC oil at prices then current would impose a bill of 2.5 billion dollars on Eastern Europe.[123] He suggested that a limit of 20 million tonnes could be expected and that this would necessitate the negotiation of substantial trading credits on the open market.[124]

The Eurocurrency market was growing rapidly and loan demand on Western Europe stagnated somewhat after 1970. This was even before the onset of the economic recession which was in turn exacerbated by the oil supply and price rises of 1973-1974. It became evident that Comecon could absorb Eurofinance at a time when its need to fund the import of raw materials and machinery was increasing. Thus the possibility of alleviating short-term supply tensions in fuels, through low-cost loans from Euromarkets emerged as an option in Comecon energy planning.

References

1. Yu S Shiryaev, S M Iovchuk (eds), *Proizvodstvennaya integratsiya stran-chlenov SEV*, Moscow, Nauka, 1972, pp 129-130.

2. H Ufer, 'Wachstums- und Strukturprobleme der Energiewirtschaften der RGW-Länder', *Wirtschaftswissenschaft*, 8/1975, p 1128.

3. I D Kozlov, E K Shmakova, *Sotrudnichestvo stran-chlenov SEV v energetike*, Moscow, Nauka, 1973, pp 14-15.

4. I V Popov, 'Sotrudnichestvo sotsialisticheskikh stran v oblasti energetiki', *Voprosy ekonomiki*, 10/1963, p 111.

5. Kozlov, Shmakova, *op cit*, p 29.

6. V P Maksakovskii, *Toplivnaya promyshlennost' sotsialisticheskikh stran Evropy*, Moscow, Nedra, 1975, p 22. (The balance of 6.2 per cent [9.3 billion tonnes] is the Yugoslav share.)

7. (i) A Alekseev, Yu Savenko, 'Ekonomicheskaya integratsiya v razvitii toplivno-energeticheskikh otraslei stran-chlenov SEV', *Voprosy ekonomiki*, 12/1971, pp 49-52; (ii) S M Lisichkin, *Energeticheskie resursy i neftegazovaya promyshlennost' mira*, Moscow, Nedra, 1974, pp 58, 64, 70; (iii) V P Maksakovskii, *op cit*, pp 23-26.

8. V P Maksakovskii, *Toplivnye resursy sotsialisticheskikh stran Evropy*, Moscow, Nedra, 1968, p 10.

9. E Taeschner, 'Tendenzen und Probleme der Energiewirtschaft der RGW-Länder', *Zeitschrift für den Erdkundeunterricht*, 6/1970, p 218.

10. *ibid*, p 219.

11. W Siegert, 'Die wirtschaftliche Bedeutung des Erdöls sowie Fragen des Erdölbedarfs und der Erdölbereitstellung in den sozialistischen Ländern Europas', *Energieanwendung*, 7/1969, p 168.

12. *ibid.*

13. Kozlov, Shmakova, *op cit*, p 79.

14. A Zubkov, 'Osobennosti mezhdunarodnoi kontsentratsii investitsii pri reshenii toplivno-syr'evoi problemy stran SEV', *Voprosy ekonomiki*, 9/1972, p 77.

15. Ufer, *op cit*, p 1134-1135, 1137.

16. (i) A I Zubkov, 'SSSR i reshenie toplivno-energeticheskoi i syr'evoi problemy v stranakh SEV', *Istoriya SSSR*, Jan/Feb 1976, p 58; (ii) N I Zakhmatov, S S Yakushin, 'Sotrudnichestvo stran-chlenov SEV v reshenii toplivno-syr'evykh problem', *Izvestiya AN SSSR Ser. ekon.*, 4/1976, p 158.

17. Zakhmatov, Yakushin, *op cit*, p 156.

18. *ibid*, pp 157-158.

19. *ibid*, p 158.

20. J G Polach, 'The Development of Energy in East Europe', in *Economic Problems in Countries of Eastern Europe*, Washington DC, US Congress, Joint Economic Committee, 1970, p 411.

21. *ibid*, p 412.

22. S Wasowski, 'The Fuel Situation in Eastern Europe', *Soviet Studies*, July 1969, pp 35-51.

23. *ibid*, p 41.

24. *ibid*, pp 50-51.

25. W Bröll, 'Die energetische Integration des RGW-Raumes', *Osteuropa Wirtschaft*, March 1968, pp 26-49.

26. *ibid*, p 35.

27. *ibid*, pp 37-39.

28. S Baufeldt, 'Die künftige Erdöllucke im RGW vor dem Hintergrund des sowjetischen Engagements in Nah-Mittel-Ost', *Osteuropa Wirtschaft*, June 1973, pp 35-54.;

29. *ibid*, p 44.

30. *ibid*, p 45.

31. K Ogawa, 'Economic Conditions in Eastern Europe and Energy Problem', *Chemical Economy and Engineering Review*, November 1975, pp 12-18.

32. *ibid*, pp 14-15.

33. *ibid*, p 18.

34. *ibid*.

35. B Korda, I Moravcik, 'The Energy Problem in Eastern Europe and the Soviet Union', *Canadian Slavonic Papers*, 3/1976, pp 1-14.

36. *ibid*, p 2.

37. *ibid*, pp 6-7.

38. *ibid*, p 14.

39. *ibid*, pp 9-14.

40. Maksakovskii, *Toplivnaya promyshlennost'*. . ., (1975), pp 139-140.

41. Note that Lisichkin's classification of reserves might not conform to the Soviet and Eastern European standard. He uses the term 'validated' ('dostoverennye'), whereas the standard term is 'proven/explored' ('razvedannye'). It is likely that Lisichkin's figures include reserves in the C2 category.

42. Lisichkin, *op cit*, pp 63, 93, 103.

43. Maksakovskii, *Toplivnaya promyshlennost'*. . ., (1975), p 129.

44. Lisichkin, *op cit*, p 63.

45. *ibid*, p 70.

46. Maksakovskii, *Toplivnaya promyshlennost'*. . ., (1975), pp 179, 189, 191, 195.

47. Polach, *op cit*, p 379.

48. T Khristov, 'Novi tendentsii v razvitieto na energetikata v Bolgaria', *Geografiya (Sofia)*, 8/1970, p 1.

49. *ibid*, p 2.

50. (i) Maksakovskii, *Toplivnye resursy . . .*, (1968), pp 135-137; (ii) Lisichkin, *op cit*, pp 60-61. The two major crude oils available in Bulgaria, Tyulen and Dolno-Dybin, balance each other conveniently. Tyulen is a heavy, viscous oil of low sulphur content, and used primarily for the production of fuel oil, bitumen and lubricants. Dolno-Dybin is light and yields gasoline, kerosene, middle distillates and naphtha. See V Marinov, 'Razvitie na nashata neftodobivna i neftoprerabotvashcha promishlenost', *Planovo stopanstvo*, 7/1968, p 56.

51. K F Schappelwein, 'Die Energiewirtschaft der VR Bulgarien', *Osteuropa Wirtschaft*, 1/1974, p 52.

52. Marinov, *op cit*, pp 78-80.

53. G Pavlov, 'Gazovaya promyshlennost' Bolgarii', *Gazovaya promyshlennost'*, 5/1970, p 7.

54. *ibid*, p 8.

55. *ibid*, p 9.

56. I Kondov et al, 'Razshiryavaneto na energiinata baza i rastezhot na neinata efektivnost', *Planovo stopanstvo*, 2/1976, p 36.

57. *ibid*, p 37.

58. Lisichkin, *op cit*, p 65.

59. *ibid*, p 66.

60. Maksakovskii, *Toplivnaya promyshlennost' . . .*, (1975), pp 165-166.

61. V Bese, 'Hungary's Mineral Oil and Gas Industry', *Marketing in Hungary*, 4/1971, p 10.

62. Lisichkin, *op cit*, p 67.

63. *Népszabadság*, 10 January 1974, p 1.

64. *Népszabadság*, 8 February 1974, p 4.

65. *Népszabadság*, 19 March 1975, pp 6-7.

66. Maksakovskii, *Toplivnaya promyshlennost' . . .*, (1975), pp 150-151.

67. T Garai, 'Razvitie gazovoi promyshlennosti Vengrii', *Gazovaya promyshlennost'*, 5/1970, p 10.

68. Maksakovskii, *Toplivnaya promyshlennost' . . .*, (1975), pp 193-194.

69. *Marketing in Hungary*, 3/1977, p 2.

70. Lisichkin, *op cit*, p 69.

71. H Wambutt, 'Planmässige Entwicklung der Energiewirtschaft der DDR', *Einheit*, 6/1974, p 706. (The international average is 60-80 per cent.)

72. K Siebold, 'Die Aufgaben der Energiewirtschaft der DDR bei der Erfüllung der Hauptaufgabe des Fünfjahrplanes', *Energietechnik*, 12/1972, p 534.

73. *Petroleum Press Service*, September 1973, p 337.

74. Lisichkin, *op cit*, p 71.

75. Maksakovskii, *Toplivnaya promyshlennost' . . .*, (1975), p 187.

76. Wambutt, *op cit*, pp 706-707.

77. G Schirmer, J Wartenberg, 'Zur Rolle der Energiewirtschaft im volkswirtschaftlichen Reproduktionsprozess – einige ausgewählte Probleme', *Wirtschaftswissenschaft*, 4/1975, p 540.

78. Wambutt, *op cit*, p 710.

79. K Siebold, 'Effektiver Einsatz der Investitionen – ein entscheidender Faktor bei der Intensivierung der Kohle – und Energiewirtschaft', *Energietechnik*, 7/1975, p 294.

80. Lisichkin, *op cit*, p 87.

81. Maksakovskii, *Toplivnaya promyshlennost'*. . ., (1975), pp 139-140.

82. *ibid*, p 151.

83. *ibid*, p 152.

84. Ya Zhytka, 'Perspektivy razvitiya gazovoi promyshlennosti Pol'shi', *Gazovaya promyshlennost'*, 5/1970, p 20.

85. Lisichkin, *op cit*, pp 86, 88. (Note that the figure for energy consumption in 1975 is given as 183 million tsf on p 86. This is probably a misprint: the figure of 146 Mtsf is given on p 88, where the figure of 183 Mtsf is shown as being that for 1980.)

86. J Molenda, 'Rola gazu ziemnego w gospodarce Polski', *Wiadomości naftowe*, 10/1973, p 218.

87. 'Prospects for Polish Natural Gas 1975-1980', *Gas World*, July 1974, pp 384-385.

88. Molenda, *op cit*, p 221.

89. *Słowo powszechne*, 27 November 1975, p 2.

90. Lisichkin, *op cit*, p 92.

91. L Bednarz, 'Nowe perspektywy rozwoju przerobki ropy naftowej w Polsce', *Nafta (Krakow)*, 5/1973, p 193.

92. Lisichkin, *op cit*, p 93.

93. *ibid*, p 94. Maksakovskii, *Toplivnaya promyshlennost'*. . ., (1975), pp 33-34. The early development of the Romanian oil industry is covered in detail in M Pearton, *Oil and the Romanian State 1895-1948*, Oxford, Clarendon Press, 1971. A brief analysis of the development of the Romanian gas industry is given by F W Carter, 'Natural Gas in Romania', *Geography*, April 1970, pp 214-220.

94. G Pacoste, 'Producţia şi utilizarea ţiţeiului şi a gazelor naturale in România', *Energetica (Bucharest)*, 6/1971, p 256.

95. Lisichkin, *op cit*, p 93. Maksakovskii, *Toplivnaya promyshlennost'*. . ., (1975), p 129.

96. Maksakovskii, *Toplivnaya promyshlennost'*. . ., (1975), p 195.

97. *Petroleum Economist*, February 1977, p 73.

98. (i) Carter, *op cit*, p 218; (ii) M Valais, M Durand, *L'Industrie du Gaz dans le Monde*, Paris, Editions Technip, 1975, (2nd edit.), pp 1-63; (iii) Maksakovskii, *Toplivnaya promyshlennost'*. . ., (1975), p 195.

99. Lisichkin, *op cit*, p 94.

100. G Pacoste, G Aldea, 'Preocupări pentru creşterea factorului final de recuperare la zăcămintele de ţiţei din RS România', *Mine, Petrole şi Gaze*, 4/1975, pp 168-169.

101. Maksakovskii, *Toplivnaya promyshlennost'*. . ., (1975), p 140.

102. *ibid*, p 145.

103. *Scînteia*, 3 August 1974, p 2.

104. Pacoste, *Producţia şi utilizarea* . . ., (1971), p 258.

105. B Popa, T Mihailescu, 'Aspecte din dezvoltarea exploătarii gazului metan în Romania', *Petrol şi gaze*, 1/1972, p 59.

106. *Statisticheskii ezhegodnik stran-chlenov SEV*, 1976, p 78.

107. V Boescu, 'Consumuri cît mai reduse de combustibili şi energie pe unitatea de produs', *Revista economică*, 12/1975, pp 9-10.

108. Maksakovskii, *Toplivnaya promyshlennost'*. . ., (1975), p 150.

109. The Romanian government has gone so far as to enter into a preliminary agreement with the American Occidental Petroleum Company Ltd, to invest $50 million in developing a high grade coal mine in Virginia USA. Possible output is 1 million tonnes per year by 1980, by which time Romania would be entitled to one-third of the output, with the option of buying a further one-third. *Petroleum Economist*, August 1975, p 313.

110. Lisichkin, *op cit*, p 102.

111. *ibid*, p 110.

112. Kozlov, Shmakova, *op cit*, p 25.

113. Maksakovskii, *Toplivnaya promyshlennost'*. . ., (1975), p 153.

114. *ibid*, pp 164-165.

115. *ibid*, p 189.

116. B Korda, 'Economic Policies and the Energy Pinch in Czechoslovakia', *The ACES Bulletin*, Vol XLVII No 2/3 (Winter 1975), pp 3-26.

117. *ibid*, p 3.

118. *ibid*, p 8.

119. J Kures in *Hospodářské noviny*, 50/1974, p 15.

120. Korda, *op cit*, p 21.

121. K Houdek, 'Uspory energie jako předpoklad rozvoje národního hospodářství', *Plánované hospodářství*, 8/1974, pp 27-28.

122. Korda, Moravcik, *op cit*, pp 13-14.

123. J R Lee, 'Petroleum Supply Problems in Eastern Europe', in *Reorientation and Commercial Relations of the Economies of Eastern Europe*, Washington DC, US Congress, Joint Economic Committee, 1974, p 417.

124. *ibid*.

Chapter 5

The Comecon Oil and Gas Industries, 1976-1980

The energy economies of the Soviet Union and Eastern Europe entered a period of substantial change in the course of their 1971-1975 Plans. The Soviet Union maintained the discovery rate of reserves of coal and gas but reserves of oil, although believed to be extensive, were not being discovered or upgraded sufficiently rapidly in the categories A, B and C1. Reserves of energy resources in these categories are now located predominantly in the Arctic North and in Siberia, and almost all growth in production in the 1976 to 1980 period was scheduled for these areas. This poses severe problems in exploration, transportation and storage due to the inhospitable nature of the terrain and climate. In 1975, the Soviet Union exported 130 million tonnes of crude oil and refined products, and the trend over the period from 1971 to 1975 showed a steady increase in deliveries to Comecon and the world market. Large-scale export contracts for natural gas were negotiated during this period and the build-up of gas trade became visible in the last two years of the ninth Plan. The Soviet Union benefited from the increased world market price for oil in the aftermath of the events of 1973 and 1974, when there seemed to be some doubt in the West as to the future of Soviet production and export capacity. Soviet energy consumption grew at some 5 per cent per year during the ninth Plan and the major part of the increase in demand was met mostly by gas but also by oil.

Eastern European countries reoriented their energy economies away from the dominant reliance on coal to imported oil and gas at a time when the purchase price from the Soviet Union and from other suppliers had increased substantially. Though prices for oil and gas imports from the Soviet Union are still below the world prices, they are moving closer to the latter under the terms of a revised pricing formula for intra-Comecon deliveries of raw materials negotiated in 1975. Total energy consumption in Eastern Europe grew by some 4 per cent from 1971 to 1975, but within this trend the oil and gas requirements grew at a rate of 10 per cent.

The demands imposed on the energy industries relate to the rate of economic development of Comecon as a whole. A striking feature of the Soviet Plan for 1976-1980 is the general slowdown in rates of growth in most sectors. The industrial target growth rates are the lowest recorded, but this is counterbalanced by ambitious (and necessary) plans for recovery in the agricultural sector. The general direction of the Soviet Plan for 1976-1980 is given in table 5.1, and of the fuels industries in table 5.2.

In Eastern Europe the 1971-1975 Plans for growth of net material product

Table 5.1 The Framework of the Ninth and Tenth Soviet Five-Year Plans

Sector	*1971-1975 Plan (1970 = 100)*		*1976-1980 Plan (1975 = 100)*
	Plan	*Actual*	
National Income	138.6	128	126
Industrial Output	147.0	143	137
Producer Goods	146.3	145	140
Consumer Goods	148.6	138	131
Agricultural Output	121.7*	113*	116*
Total Investment	141.6*	139*	125*

**Quinquennium as % of previous quinquennium.*

Sources: A Nove, *The Times*, 5 January 1976, p 14; J D Park, R A Clarke, *ABSEES (Special Section)*, July 1976, p (i)

Table 5.2 Direction of Soviet Fuel and Energy Production 1970 to 1980

	1970	*1975P*	*1975A*	*1980P*	*1975A as % of 1970*	*1980P as % of 1975A*
Electricity (billion kWh)	740	1,065	1,038	1,340-1,380	140.2	131.0*
Oil and Gas Condensate (million tonnes)	353	505	491	620-640	139.1	128.3*
Natural Gas (billion cubic metres)	198	320	289	400-435	144.5	144.6*
Coal (million tonnes)	624	695	701	790-810	112.3	114.1*

**Percentage calculations based on mid-point of range; P = Plan; A = Actual.*

Source: Park, Clarke, *op cit*, p (ii)

were fulfilled or exceeded. An exception was Bulgaria where an underfulfilment of the aggregate industrial Plan was recorded. However, domestic production of energy fell short of Plan targets throughout Eastern Europe and overall growth rates were below those attained in the sixties. An actual decline in production of oil and gas was experienced in some countries, and this resulted in a greater utilization of coal than was initially envisaged so that deliveries of oil and gas feedstocks to the petrochemical industry could be maintained. During 1975, however, the annual rate of growth in all Comecon countries showed a decline over the previous year: the growth rate of aggregate net material product in Eastern Europe declined from 8.3 per cent in 1974 to 7 per cent in 1975. The major indicators of economic development in Eastern Europe for 1971 to 1975 together with targets for 1980 are detailed in table 5.3.

The need for energy conservation is highlighted in the Plans of all countries, reflecting the poor prospects for increased indigenous production of energy

Table 5.3 Direction of Eastern European Economic Growth, 1971-1975 and 1980 Plan
(Average annual growth rates in %)

	Bulgaria	*Czechoslovakia*	*GDR*	*Poland*	*Romania*	*Hungary*	*Eastern Europe*
Net Material Product							
1971-1975 Plan	7.7-8.5	5.1	4.9	7.0	11.0-12.0	5.5-6.0	6.8
1971-1975 Actual	7.9	5.7	5.4	9.8	11.3	6.2	7.8
1976-1980 Plan	8.2-8.7	4.9-5.2	4.9-5.4	7.0-7.3	11.0	5.4-5.7	6.7
Industrial Production							
1971-1975 Plan	9.2-9.9	6.0	6.0	8.5	11.0-12.0	5.7-6.0	7.7
1971-1975 Actual	9.1	6.7	6.4	10.7	13.1	6.5	8.7
1976-1980 Plan	9.2-9.9	5.7-6.0	6.0-6.4	8.2-8.5	11.2	5.9-6.2	7.5
Foreign Trade Turnover							
1971-1975 Plan	9.8-10.5	6.4-6.6	8.0	9.5	10.0-11.5	7.0-8.5	9.0
1971-1975 Actual	16.4	12.1	13.2	22.1	15.6	14.9	15.8
1976-1980 Plan	9.9-10.5	6.2-6.5	9.7	8.5	12.4	7.7-8.5	9.0

Source: A Askanas, H Askanas, F Levcik, 'Die Wirtschaft der RGW-Länder 1971-1975 und die geplante Entwicklung bis 1980', *Monatsbericht*, 3/1976, Osterreichisches Institut für Wirtschaftsforschung.

throughout the bloc. In view of the enhanced opportunities for earnings from the export of fuels and the increasing need of the Soviet Union and Eastern Europe to import Western technology, energy conservation might have become a feature of economic development in Comecon even without the production difficulties experienced between 1971 and 1975.

The American analyst M Slocum, writing at the end of 1974, went so far as to state that 'the energy situation in the Soviet Union poses a serious dilemma: in order to sustain planned economic growth, the Soviet Union will have to increase progressively the imports of oil and gas from the Middle East area, will have to import technology, and obtain immense investment credits'.[1] She went on to contend that 'the demand for fuel and energy has been greater than supply for a number of years',[2] and that this has resulted in 'skyrocketing energy shortages in . . . industrialized areas'.[3] Moreover the difficulties encountered by the Soviet Union in energy development are said to have prompted the Soviet executive into 'trying desperately to penetrate Western trade barriers', which would 'unquestionably give the West added bargaining power in the negotiation of trade agreements with the Soviets and provide a basis for requiring assurance of alternative modes for the repayment of loans by the Soviet Union'.[4]

A somewhat different and less categorical analysis was written in 1976 by the British economist Philip Hanson.[5] The object of his enquiry was to discover why 'in the past few years . . . the Soviet energy balance has begun to seem precarious'.[6] He stressed that towards the end of the ninth Soviet Plan, energy policy shifted to favour the rehabilitation of coal, facilitated by a substantial development effort in the Ekibastuz and Kansk-Achinsk coalfields. This policy was also to include a nuclear power station construction program for European Russia and the Ural region that provided for a nuclear component of 13-15 million kilowatts out of a total of 67-70 million kilowatts scheduled for completion during the tenth Plan.[7] Philip Hanson coupled this policy with a general slowdown in rates of economic growth in the tenth Plan compared with the ninth as calculated to ease the pressure that had built up in the oil sector. His conclusion was that the planned increase of energy supply was adequate to sustain the economic growth rates in the tenth Plan. He argued that there would, however, be a shift to less efficient fuels, incurrence of very high costs in developing Siberia and perhaps the import of significant (unquantified) amounts of Middle East oil, but that the energy problems of the Soviet Union and the West 'are not sufficiently awful to ensure peace and goodwill between Moscow and Washington'.[8]

A further Western perspective was that of the American analysts E E Jack, J R Lee and H H Lent.[9] They pointed out that the Soviet Union was the only industrialized nation in the world that was self-sufficient in energy and that this position was likely to be maintained 'for the foreseeable future'.[10] These authors felt that the 1980 production target for coal could probably be met without great difficulty, whereas the 'ambitious' targets for oil and gas production were not likely to be achieved. This would then result in a shortfall of two to five per cent of planned energy production, necessitating the imposition of energy conservation measures or an adjustment in foreign trade.[11] Alternatively they suggested that in the event of lags in the Soviet economy as a whole the overall supply and demand balance in the energy sector could be maintained.[12]

A report of the American Central Intelligence Agency concerned specifically with future oil production was published in April 1977 and updated in July of that year.[13] The report concluded that as a result of technological backwardness the point at which the Soviet Union could prove sufficient reserves to offset depletion and to add to existing production has been passed, and that production of oil and gas condensate would peak 'possibly as early as next year (1978) and certainly not later than the early 1980s'.[14] This, the report argued, would result in the Soviet Union's entry into the world petroleum market as a competitive purchaser of OPEC oil, since the effect of fuel substitution would be minimal in the short term.[15]

This study has been hotly disputed in a number of quarters, and most notably by Soviet specialists. The report correctly stressed the considerable technical problems that beset the Soviet oil industry. In particular, the amount of new capacity that had to be brought on stream simply to offset depletion was cited as a problem of increasing severity, as was the declining success rate recorded in exploratory drilling. Whilst there is ample evidence to support these statements, the conclusions of the report are overdrawn and lack an international comparison to serve as a yardstick. The technical delays experienced in Alaska, for example, illustrate only too well the problems that the Soviet Union is currently facing on a greater scale in West Siberia. In short, the report identified a series of genuine technical problems besetting the Soviet oil industry and assumed that none will be solved. It also ignored the substantial adjustments that Soviet planners have made in the energy balance to allow for the oil industry's problems.

It will be seen from the aforementioned analyses that a considerable range of opinion exists on the prospects for Soviet energy development in the tenth Plan, not only in relation to growth targets but also to future trade patterns. The question of trade is dealt with in Chapter 6.

A detailed study of the methodology of energy planning in the Soviet Union was produced in 1973 by the Energy Institute of the Siberian Division of the Academy of Sciences in Novosibirsk.[16] Though the work was concerned primarily with alternative methods of energy analysis and planning, involving the use of complex linear programming techniques, the authors do offer some perspective on the short-term trend in the Soviet energy balance. They pointed out that there was a substantial difference between energy demand in European Russia and the Eastern regions. The former accounted for most of the hydrocarbon consumption: the share of coal in the European area's energy balance in 1970 was 20 per cent and this is expected to fall to 5-7 per cent by the end of the century, assuming successful development of the nuclear power program. In contrast, in Siberia and the Trans-Baikal areas 60 per cent of energy consumption is in the form of coal and a further 10-15 per cent in fuelwood, and this situation is likely to be maintained. The authors argued that smaller consumers would utilize natural gas, distillate fuels and fuel oil, since they did not enjoy the same economies of scale in conversion of coal as, for example, power stations. In the long (unspecified) term, the authors stated, oil and gas would be produced from deposits in the Lena-Vilyuy area and the Irkutsk oblast', that is to say, they viewed the development of East Siberian oil and gas as essential, to support the long-term development plans envisaged during the ninth Five-Year Plan.[17]

A later Soviet analysis of the energy balance pointed to an intrinsic tension between supply and consumption, and outlined the extent to which conservation measures in electricity generation have contributed to a more rational use of fuel.[18] On a different plane it was suggested that attempts to determine an optimal energy balance were subject to a number of factors that were not only difficult to predict but which also exhibited a tendency to fluctuate, such as export prices. The criterion of comparative initial financial outlay must also be tempered by consideration of the effect of change on the demand for individual fuels in the various branches of the economy.[19] This analysis concluded that in view of the tightening balance between supply and demand the prime objective should be that of securing maximum flexibility in the choice of fuel amongst the larger users. Power stations are an example and to the extent that a short-term interfuel switch becomes possible, this policy could be initiated in the Volga-Ural region, where coal, natural gas and fuel oil were freely available.[20]

However, despite the Soviet planners' preference for securing as flexible an energy policy as possible the existing technical scope is limited when the economic dimension is considered. Soviet planners bear in mind that whereas the long-term delivered cost of Siberian oil and gas is expected to be lower than alternative conventional fuels in European Russia,[21] the potential for export earnings, primarily in Western Europe, has had a substantial influence on the rate at which Siberian resources have been developed. The reason is that long-term policy is to depress demand for conventional fuel in European Russia through the expansion of nuclear power.[22] Despite the relatively minor role assigned to nuclear power in the short term, assumptions concerning the level and timing of its contribution have a marked influence on short-term decisions concerning production and utilization of conventional fuels. It is likely that the effect of nuclear development in the Soviet Union will not be felt before the mid-eighties, when nuclear power could account for 25 per cent of electricity produced.[23] Therefore a policy of relatively rapid depletion of coal, and to a lesser extent of gas reserves can be undertaken. However, the high initial cost of such a policy can only be offset by increased assistance, human, material and financial, from interested outsiders, fellow members of Comecon, Japan and the US.

Oil Developments in the Soviet Union 1976-1980

The initial target for oil production in the Soviet Union in 1980, as given in the basic outlines for the tenth Five-Year Plan released in December 1975, was put at 620 to 640 million tonnes.[24] The final target was subsequently confirmed at 640 million.[25] Other data relating to the oil industry indicated that time spent in proving new wells was to be cut by 25 to 30 per cent, utilization of wellhead gas raised to 43-45 billion cubic metres and primary refining capacity increased by 25 to 30 per cent, with increasing emphasis on the production of high octane gasoline, aviation fuel and low-sulphur distillates.[26]

The target for oil production in West Siberia in 1980 was set at 300 to 310 million tonnes,[27] which means that Siberia would account for the total net increase in production. At the outset of the tenth Plan the then Soviet Oil

Minister, V D Shashin, called for a substantial transfer of exploratory effort to West Siberia and the Baltic and Caspian offshore areas.[28] He pointed out that during the ninth Plan the Samotlor oilfield in West Siberia accounted for an average increase in production of 16 million tonnes and that a corresponding figure of eight million was anticipated during the tenth Plan.[29] Hence the discovery of Samotlor-size fields would be necessary in order to offset depletion. To complement production developments in West Siberia, the Timan-Pechora area of the Komi ASSR was scheduled to contribute some 25 million tonnes in 1980, and in the Perm oblast' a 1980 production level of 30 million tonnes was foreseen, this latter figure representing a 7 million tonne increase on 1975.[30]

The role of oil in the energy balance was reappraised in the light of problems encountered during the ninth Plan. Since reserves of coal exceed those of oil and to a lesser extent of gas, it was decided to give preference to the use of coal as a boiler and furnace fuel. Oil would be saved for maximum production of gasoline and naphtha, and gas for use as fuel in highly critical processes, and as petrochemical feedstock. The share of oil in the energy balance was intended to remain at the level reached in 1975, perhaps even showing a slight fall, through to 1980.[31] Conservation of hydrocarbon fuels thus became a guiding principle.

Whereas the average annual growth rate of oil production in the Soviet Union between 1966 and 1970 had been 7.5 per cent, the rate decreased to 6.8 per cent between 1971 and 1975. To meet the final 1980 production target of 640 million tonnes an average annual growth rate of 5.4 per cent is required. Moreover the rate of growth required to meet the target for 1976 was 5.9 per cent on the 1975 figure. Lower annual growth rates were therefore anticipated by the end of the tenth Plan.

The problem of depletion in oilfields located in the Western areas has complicated the planning process, in that West Siberia is now being developed at a rate well in excess of that which had been previously judged optimal. Towards the end of the ninth Plan Soviet geologists forecast an optimal production level of 230 to 260 million tonnes per year from West Siberian operations by 1980, and oil industry planners gave their estimates of optimal levels as 270 million tonnes in 1980, rising to 330 in 1990 and to 360-380 by the year 2000.[32] The target of 310 million tonnes finally adopted for 1980 has long-term implications for production which require action during the tenth Plan. It is thought that growth rates for West Siberia in the tenth Plan, if achieved or exceeded, are unlikely to be scaled down subsequently. Therefore increased exploratory effort will be required up to 1980 in order to ensure the continued capacity of West Siberian operations to provide the higher production levels likely to be planned in the eighties.[33] The implication of the target set for West Siberia in the tenth Plan is that a production level of 500 million tonnes per year will be reached by 1990 with a consequent and commensurate requirement for logistic development.[34]

The dislocating effect of the rapid development of West Siberia is well illustrated in the case of the Samotlor oilfield. The production target originally envisaged for 1975 was 60.6 million tonnes: however, as a result of production shortfalls elsewhere in the Soviet Union during the ninth Plan the final level in Samotlor was 86.5 million tonnes.[35] This 34 per cent overfulfilment was reflected in a capital requirement some 12 per cent above Plan and necessitated the application of temporary measures to boost production, including a higher level of water-flooding than anticipated, which resulted in a deteriorating ratio of recovered oil to water.[36] The rate of depletion of the Samotlor oilfield is one of the most decisive factors influencing plans for production beyond 1980. Newly discovered fields in West Siberia are so far proving somewhat smaller than the Samotlor field, and it is estimated that in the course of the tenth Plan it will be necessary to bring on stream about seven new fields per year, including the discovery of at least one field of the size of Samotlor. In the 10 years to 1975 a total of only 16 new fields of major significance were brought on stream.[37] The scale of the problem is such that approximately two thirds of the target increase in production capacity of 100 million tonnes per year between 1976 and 1980 will serve to offset the effects of depletion in existing operations.[38]

This fact has prompted Soviet planners to consider more seriously the production potential of East Siberia. Whilst small quantities of oil and gas have been produced in East Siberia since the early sixties, the true potential of the region is unknown: however, geologists are apparently more confident of finding further oilfields of the size of Samotlor in East rather than in West Siberia.[39] It is thought that offshore exploratory activity in the Baltic and Caspian Seas, stated to be a priority for the tenth Plan, will not begin to provide additional quantities above current levels, before 1980. Lack of appropriate drilling equipment is the principal problem. For example, in Caspian offshore operations drilling equipment is adequate to probe only up to two thousand metres below the sea bed, whereas the major oil deposits are to be found between two and three thousand metres. This has meant that approximately half of the exploratory wells sunk so far have had to be abandoned due to the inadequacy of production technology.[40] The extent of the problems faced in offshore drilling technology is highlighted by consideration of the efforts devoted to extracting oil from wells previously abandoned as uneconomic. A novel method of oil extraction has been developed in order to rework some of the oilfields in the Baku area. This process involves the excavation of underground caverns and the installation of drilling and pumping equipment so that crude oil can be 'mined', stored temporarily underground and then pumped to the surface through conventional pipelines.[41] Further examples of the reworking of abandoned wells are to be found in the Caucasus area, Central Asia and Kazakhstan.[42]

The question of exploration and utilization of gas condensate is recognized as being of increasing importance during the tenth Plan. When the Urengoi field comes into full production in 1978 substantial amounts of gas condensate will be available. At the moment its role is not fully defined by Soviet planners. During the ninth Plan production remained at a relatively low level and this was not increased towards the end of the Plan to counteract underperformance in the oil and gas industries. The immediate benefit of gas condensate is that apart from its high calorific value it needs little processing before use either in the manufacture of gasoline or as petrochemical feedstock, and it requires a considerably lower investment than that needed to derive the same quantity of final product from crude oil.[43] The availability of condensate will, to some extent alleviate the pressure on oil demand towards the end of the tenth Plan.

The oil production plan for 1976-1980 is outlined in table 5.4, and the

Table 5.4 Soviet Oil Production 1975 and Plan for 1976-1980
(million tonnes, including gas condensate)

	1975A	*1976P*	*1976A*	*1977P*	*1977A*	*1978P*	*1980P*	*1980 Final Plan*
USSR Production	491[1]	520[3]	520[5]	550[7]	546[9]	575[11]	620-[13] 640	640[15]
West Siberian Production	148[2]	180[4]	184[6]	215[8]	220[10]	250[12]	300-[14] 310	310[8]
West Siberia as % of Total Production	30.1	34.6	35.4	39.1	40.3	43.5	46.8-50.0	48.4

P = Plan; A = Actual

Sources: 1. *Pravda*, 1 February 1976, p 1; 2. *ibid*, p 2; 3. *Izvestiya*, 3 December 1975, p 2; 4. *ibid*, p 3; 5. *Neftyanoe khozyaistvo*, 3/1977, p 3; 6. *ibid*, p 4; 7. *Ekonomicheskaya gazeta*, 8/1977, p 12; 8. *Ekonomicheskaya gazeta*, 16/1977, p 12; 9. *Pravda*, 28 January 1978, p 1; 10. Estimate; 11. *Pravda*, 15 December 1977, p 2; 12. *ibid*, p 3; 13. *Izvestiya*, 14 December 1975, p 2; 14. *ibid*, p 5; 15. *Izvestiya*, 28 October 1976, p 3

growing importance of West Siberia is evident.[44] It is recorded that West Siberia produced some three million tonnes above its target for 1976, the Samotlor oilfield accounting for some 110 million tonnes of total West Siberian production.[45] The net increase in production in 1976 compared with 1975 was 5.9 per cent and the Plan for 1977 called for a further increase of 5.7 per cent, showing a steady production increase and as anticipated a slightly declining growth rate towards 1980. Production in certain older fields continued to show an expected decline.

In addition to the overfulfilment of the Plan by the Tyumen' oil and gas production association (Glavtyumenneftegaz), the collectives in Perm, the Komi ASSR, Udmurtia and Georgia recorded production in excess of Plan. Operations in Tataria, Bashkiria, the Kuybyshev oblast', Groznyi, Orenburg, Stavropol and the Emba krai performed on target for 1976: however, shortfalls were recorded in Mangyshlak, Belorussia, Azov and Caspian offshore operations, Turkmenia, the Ukraine and the Lower Volga.[46] Table 5.5 details the 1976 production of major associations.

Table 5.5 1976 Production of Oil by Major Production Association (million tonnes, including gas condensate)

Association	*Production*
Glavtyumenneftegaz	214.5
Tatneft'	98.9
Bashneft'	40.0
Kuybyshevneft'	31.7
Permneft'	24.7
Mangyshlakneft'	19.5
Turkmenneft'	13.9
Orenburgneft'	12.9
Kaspmorneft'	11.4
Komineft'	10.9

Source: *Ekonomicheskaya gazeta*, 8/1977, p 2

During 1976 some progress was made in developing the utilization of wellhead gas. A further 3 billion cubic metres per year of processing capacity was installed, fulfilling the target set and enabling utilization to be increased by 2 billion cubic metres. However, it is noted that wastage of 17 billion cubic metres in 1976 was still regarded as excessive.[47]

The annual Plan for 1977 called for an increase in production of oil and condensate to 550 million tonnes: the figure achieved was 545 million.[48] Again the major contributor to the increase was West Siberia and there was little prospect of substantial additional contribution from other producing areas. Particularly severe problems were caused by delays in implementing the program of offshore development. This was notable at the Mangyshlak peninsula in the Caspian Sea, where the prospects for bringing on stream substantial oilfields were thought by Soviet geologists and economists to be attractive, judging from exploratory evaluation carried out during the ninth Plan.[49] Despite the apparent attractiveness of Soviet offshore oil deposits, output declined during the ninth

Plan.[50] Soviet planners seek to arrest this decline and to reach a production level of 18 million tonnes in 1980. The trend in Soviet offshore oil production is outlined in table 5.6.

The performance of the Soviet oil industry in the first two years of the tenth Plan has been slightly below target and this suggests that production in 1980 might turn out to be at the lower rather than the upper end of the range of 620 to 640 million tonnes. However, this depends as much on the extent to which depletion in the Volga-Ural area leads to declining production there, as on the capacity of West Siberia to overfulfil its target. Moreover the extent to which the Construction Ministry for Oil and Gas Enterprises fulfils that part of its target relevant to the oil industry is just as important a factor influencing the success of the Oil Ministry, as their own efforts in exploration and production activity.

Table 5.6 Soviet Offshore Oil Production 1965-1980 Plan (million tonnes)

Year	*Production*
1965	11.4
1970	12.9
1971	12.5
1972	11.8
1973	12.0
1974	11.5
1975P	11.5
1980P	18.0

(converted from barrels to tonnes at rate of 7.3 barrels per tonne)

P = Plan

Source: Joseph P Riva Jr, 'Soviet Offshore Oil and Gas' in *Soviet Oceans Development*, Washington DC, US Congress, National Ocean Policy Committee 1976, p 487

Gas Developments in the Soviet Union 1976-1980

The basic directives for the tenth Plan set the 1980 production target for natural gas at 400-435 billion cubic metres.[51] Production in West Siberia was scheduled to reach 125-155 billion by that date.[52] Subsequent discussion of these figures resulted in the placing of the final target at the upper end of the scale, that is, all-Union production of 435 billion cubic metres[53] and West Siberian of 155 billion.[54] Table 5.7 outlines the pattern of planned growth in gas production from 1976 to 1980.

During the ninth Plan the average annual increase in production was 18 billion cubic metres compared with 14 billion achieved during the eighth. However, 53 billion cubic metres (58 per cent) of the total growth in production of 91 billion were obtained in the last two years.[55] The problems of overall fulfilment arose in the first two years, and related to the difficulties of coordinating production

Table 5.7 Soviet Natural Gas Production, 1975 and Plan for 1976-1980 (billion cubic metres)

	1975A	*1976P*	*1976A*	*1977P*	*1977A*	*1978P*	*1979P*	*1980P*	*1980 Final Plan*
USSR Production	289[1]	313[2]	321[3]	342[3]	346[4]	370[3]	401[3]	400-435[5]	435[3]
West Siberian Production	38[6]	46[7]	48[8]	63[9]	73[10]	97[11]	na	125-155[12]	155+[13]
West Siberia as % of Total Production	13.1	14.7	15.0	18.4	21.1	26.2	na	28.7-38.8	35.6+

na = not available; P = Plan; A = Actual

Sources: 1. *Pravda*, 1 February 1976, p 1; 2. *Izvestiya*, 3 December 1975, p 2; 3. *Ekonomicheskaya gazeta*, 6/1977, p 1; 4. *Pravda*, 28 January 1978, p 1; 5. *Pravda*, 7 March 1976, p 3; 6. *Ekonomicheskaya gazeta*, 14/1976, p 1, 22/1976, p 4; 7. *Izvestiya*, 3 December 1976, p 3; 8. *Ekonomicheskaya gazeta*, 6/1977, p 2; 9. *Ekonomicheskaya gazeta*, 16/1977, p 12; 10. Estimate; 11. *Pravda*, 15 December 1977, p 3; 12. *Pravda*, 7 March 1976, p 7; 13. *Gazovaya promyshlennost'*, 11/1976, p 2, 6/1976, p 1

and transportation, rather than of the gas industry to produce.[56] Depletion in the gasfields of European Russia was a feature of the ninth Plan, and it is likely that this will be increasingly severe throughout the tenth.[57]

The areas designated as being of prime importance in the tenth Plan were West Siberia, Turkmenia, the Orenburg oblast', the Komi ASSR and the Ukraine,[58] of which West Siberia was scheduled to provide over 30 per cent of Soviet output by 1980. West Siberian reserves are extensive and the area is expected to be the major gas-producer long after the end of the tenth Plan.[59] However, the consistent Soviet record of failure to meet production targets, with the corresponding necessity of revising them, has suggested to one Western analyst that the 1980 output of gas will, at best, be at the lower limit of the original range.[60]

According to a recent Soviet estimate West Siberia contains 70 per cent of Soviet gas reserves in all categories, and operations in the area are required to provide 85 per cent of the total increase in production during the tenth Plan.[61] The immediate benefit of West Siberian gas from the viewpoint of the production enterprises is that the bulk of reserves are concentrated in shallow deposits. Large-diameter wells sunk in the Tyumen' fields are capable of yielding 1 to 1.5 billion cubic metres of gas per 24 hours, this being five to six times the Soviet average.[62] The importance of accelerating production in the Tyumen' oblast' can be gauged from the fact that the 1976 production level of 43 billion cubic metres constituted a 10 billion increase on the 1975 level.[63] These factors are, however, conditioned by the difficulties faced in providing adequate production technology to facilitate higher output. This was most important since the discovery rate of gas reserves that were economically extractable was greater than the rate of increase in production, at least in the early part of the ninth Plan.

Though the Medvezh'e field will continue to be an important and expanding contributor to total production, the major area of interest is the Urengoi field, scheduled to reach full production in 1978.[64] There is still spare capacity in the Medvezh'e field, since failure to commission pipeline meant, that at the end of the ninth Plan, the desired production level could not be achieved.[65] The ratio of capital invested in the development of the Medvezh'e field is 40 per cent to the provision of gas production capacity and 60 per cent to infrastructural developments. Capital requirement and production costs are expected to rise in the Medvezh'e field during the tenth Plan,[66] and Soviet analysis of the prime production cost of Medvezh'e gas has established that some 55 per cent is accounted for in offsetting depreciation of capital equipment. Though it is not clear what is included in the concept of 'depreciation', the significant fact is that this percentage is considered high, and put at twice the all-Union average.[67] The major reason for this, it is thought, is that all materials and equipment have to be brought into the area: Tyumen' simply lacks an industrial base. Deficient organization of the supply function has served to generate periodic shortages and to raise costs.[68]

The Urengoi gasfield presents a different aspect in that it has extensive reserves of gas condensate in addition to natural gas. Potential production of 100 billion cubic metres per year is envisaged from this field.[69] The debate on the development of gas condensate is still in progress and the ultimate production level is still a matter for speculation. The field's contribution to production is scheduled to commence in 1978, when the Medvezh'e field will have reached its peak. It is intended that the major part of the production increase planned for Tymen' in 1978 will be provided by Urengoi operations and the oblast' is to account for 20.4 of the 23 billion cubic metre increase for the Soviet Union as a whole.[70] Between 1977 and 1979 Tyumen' operations are scheduled to contribute 70 to 75 per cent of the net Soviet increase in production and by 1980 they are to provide all of it.[71] The policy of rapid development of Tyumen' gas is intended to accelerate the process of maximum substitution of gas for fuel oil. This was initiated during the ninth Plan, principally in the Volga-Ural industrial area, and it was also intended that it should compensate for the anticipated decline in production in older fields.[72] Operations in Turkmenia, showed a substantial rate of expansion in the latter part of the ninth Plan. Some 47 of the 146 billion cubic metres produced during the Plan were obtained in 1975[73] and these operations were scheduled for further expansion up to 1980. The production target for 1976 of 57 billion cubic metres depended for its fulfilment on the continued development of the Shatlyk gasfield, of which the 1976 production target was 31 billion cubic metres.[74] Production was also to be maintained in the Naip and Gugurtli fields, which had been exploited extensively during the ninth Plan.[75] By the end of May 1976 the Shatlyk field recorded a 400 million cubic metre overfulfilment of the Plan for the first five months of 1976.[76] This in turn lifted Turkmen production to 82 million cubic metres per 24 hours and augured well for the possible provision of at least 80 billion cubic metres in 1980. This would be 20 per cent of Soviet production, and would mean that Turkmenia operations had achieved its objective.[77] Neighbouring fields in Uzbekistan, facing similar technical and logistic problems to those of Turkmenia, were scheduled to provide 36 billion cubic metres in 1976, but no increase on this level is expected to 1980.[78]

The development of the Orenburg hydrocarbon province is one of the major commitments, not only of the Soviet Union but also of each of the other Eastern European full members of Comecon, up to 1980. The gas production target for Orenburg in 1976 was set at 30 billion cubic metres, representing an increase of 9.3 billion on 1975 production.[79] Capacity to process and distribute 30 billion cubic metres per year was available at the end of the ninth Plan: however, delays in bringing Orenburg up to planned production meant that spare capacity was available at the outset of the tenth Plan.[80] These delays have been particularly costly in that this province, of all the major Soviet gas deposits is the most conveniently located to centers of consumption in European Russia. The projected production capacity in the Orenburg oblast' is 45 billion cubic metres per year to be reached in 1977-1978 when the third stage of the complex is completed.[81] On the evidence available it appears that this figure will be the likely production level through to 1980. However, the location and extent of reserves in this field are such that a higher level of production may be planned for the eighties. The increase in production planned for Orenburg in 1976 was achieved, the final figure being 31.8 billion cubic metres.[82]

The center of interest in the Komi ASSR is the Vuktyl gas and gas condensate field, which is reasonably close to consumption centers and enjoys better transport links than many other major gas-producing regions. During the ninth Plan the field produced a total of 80 billion cubic metres of gas.[83] However, it is felt that there will be short-term problems in maintaining production due to declining seam pressure and the fact that some 20 per cent of extractable reserves had been utilized by the end of 1975.[84] Gas production in the Timan-Pechora region as a whole, in which the Vuktyl field is situated, is scheduled to rise to 22 billion cubic metres in 1980.[85] It is interesting to note that the directives for gas production in the Komi ASSR in 1976 require only that the 1975 level of 17.8 billion cubic metres be maintained.[86] The medium-term significance of the Komi operation is its particularly favourable ratio of exploitable reserves to production. An estimate made in 1974 put explored reserves at 390.9 billion cubic metres, of which 367.1 are classified as 'industrial', that is, readily exploitable.[87]

The 1976 production target for the Ukraine was set at 57.5 billion cubic metres.[88] The Shebelinka field is the major contributor supplemented by a small number of new discoveries made towards the end of the ninth Plan. Maintenance of 1975 production levels is essential in the short term, since the area, which is the gathering ground for the trans-Comecon 'Bratstvo' gas pipeline, is the principal producer for export.

The regional performance of the gas industry in 1976 compared with 1975 is detailed in table 5.8. This distribution probably reflects the production balance originally desired for 1975, and constitutes an 11.1 per cent increase on 1975

Table 5.8 Soviet Gas Production by Area 1975 and 1976 (billion cubic metres)

Area	*1975*	*1976*	*1976 as % of 1975*
USSR Total	289	321	111
of which:			
Ukraine	68.2	68.7	101
Turkmenia	52.3	62.6	120
Tyumen' oblast	35.5	47.8	135
Uzbekistan	37.1	36.0	97
Orenburg oblast'	20.1	31.8	158
Komi ASSR	18.5	19.6	106

Sources: *Table 3.18*; *Ekonomicheskaya gazeta*, 6/1976, pp 1-2

production. In order to attain the target of 435 billion cubic metres in 1980, an average growth rate of 8.5 per cent is required, compared with an average annual rate of growth of 7.9 per cent actually achieved during the ninth Plan. The cumulative performance of the gas industry in the ninth Plan showed a shortfall of 71 billion cubic metres against a planned total for the five-year period of 1,290 billion. This represented a 5.5 per cent shortfall. Gas production in 1977 totalled 346 billion cubic metres,[89] which promises well for the attainment of the upper limit of the original Plan for 1980.

The target for 1977 of 342 billion cubic metres constituted an increase of

21 billion over the 1976 performance, but was 11 billion below the actual increase achieved in 1976 over 1975. This probably reflects an anticipated hiatus in growth prior to the commissioning of further capacity at Urengoi and Orenburg. It is likely that the marked increase in production in 1976 over 1975 stems from the fact that productive capacity in Tyumen', available but under-utilized at the end of the ninth Plan, was brought on stream at an early stage in the tenth. At this time, freshly injected investment together with other factors catalyzed the provision of equipment and support materials that were evidently lacking towards the end of the ninth Plan. Consequently the importance of the timing of development in Urengoi and Orenburg for the overall capacity of the Soviet Union to provide the desired increase in gas production is very great, since it is likely that any 'slack' within the interacting systems has by now been taken up. However, as in the case of the oil industry, the likelihood of the gas industry's meeting its 1980 target depends very much on the ability of other ministries to fulfil their delivery quotas of equipment and materials. Given the increased demands on Siberian operations, the difficulties of gas supply are almost certain to increase. The gas industry, like the oil industry, faces the prospect of increasing tension between supply and consumption.

Logistic Developments 1976-1980

As in the ninth Plan major decisions are tied to coping with the worsening geographical dislocation between production and consumption centers. Transport problems revolve around the installation of new pipelines and the expansion of existing systems to correspond with the growth in production planned for West Siberian operations to 1980 and beyond. The tenth Plan called for the construction of 15 thousand kilometres of trunk oil pipeline with at least a further 3,500 kilometres for the transport of refined products. Some 35 thousand kilometres of new gas pipeline were planned for the same period.[90] Apart from the problem associated with the supply of materials and their technical performance, construction engineers working in West Siberia have to contend with extremes of climate and the environment. The most severe obstacle is that some 40 per cent of Tyumen's oil- and gas-bearing area is extremely marshy in summer and frozen solid in winter, with the result that the construction 'season' is limited to three or four months, when heavy equipment may be transported.[91]

The specific importance of the logistic development of West Siberia was reflected in the creation in June 1973 of the Siberian Pipeline Construction Association (Glavsibtruboprovodstroi).[92] The principal new oil pipeline projects are those from Nizhnevartovsk to Kuybyshev and from Kholmogorsk to Surgut; in the gas sector the major projects are to commission pipelines from the Medvezh'e field via Nadym to Punga, and from Nadym via Urengoi and Punga, to join the system centered on the Vuktyl gas and condensate field.[93] The development of Siberia has important implications for the industrialization of Soviet Central Asia. Whereas there is a high level of consumption in the area of locally produced gas, demand for petroleum products is also increasing. Hence towards the end of the ninth Plan construction work was started on a major crude oil pipeline from Surgut via Omsk, Pavlodar and Chimkent, to refineries at

Fergana and Chardzhou, supplying refineries at Pavlodar and Chimkent en route.[94] The difficulties of processing oil and gas at the wellhead, under Siberian conditions, have necessitated the greater application of technology to the transportation of gas-saturated oil to processing centers located at some distance from producing fields. This technique reduces the requirement for compressor stations, since the natural pressure of the gas is utilized, and the availability of gas in oil for the further processing of both, marginally reduces the requirement for gas delivery capacity and stimulates greater usage of wellhead gas.[95]

The most important single construction project currently in progress in the Soviet Union is that of the gas pipeline from Orenburg to the Western border, on which work commenced during the ninth Plan.[96] The line will be some 2,800 kilometres long and consists for the most part of 1,420 millimetres diameter pipe.[97] The delivery capacity of the system, when fully operational in 1980, will be 28 billion cubic metres per year, of which 4 billion will be used en route by the compressor stations.[98] Once the system is running at design capacity, some two thirds of the gas will be delivered to Eastern European countries as repayment of investment credits granted to the Soviet Union. The balance of the available gas and gas condensate will be directed to consumers in European Russia.[99] Taking account of the development of the northerly hydrocarbon resources in the Komi ASSR, a pipeline from Usinsk via Ukhta and Yaroslavl' to Moscow was completed at the end of the ninth Plan. In 1976 work commenced on the new line from the Punga and Vuktyl fields to join the gathering system in the Central region, running parallel to the existing 'Northern Lights' pipeline.[100] The major achievements of 1976 were the commissioning of the trunk oil pipeline from Nizhnevartovsk to Kuybyshev, the gas pipeline from Urengoi to Ukhta and the second string of the gas pipeline from Ukhta to Torzhok.[101]

The fact of Siberia's substantial overfulfilment of the original production target for oil during the ninth Plan necessitated some rethinking of the forward requirement for pipeline capacity in the tenth Plan and beyond. In a study written in 1972 the prominent Soviet transport economist S S Ushakov put forward the view that a great movement of industrial production into eastern areas so as to lessen the net demand for energy in European Russia could not be expected in the period 1965-1985.[102] Therefore there would be a continuing need to expand the oil and gas delivery systems, and given the eastward shifting resource base, the economics of pipeline transport would become gradually more attractive.[103] The very rate at which Siberia has had to be developed serves only to underline Ushakov's conclusion. It should, however, be borne in mind that the availability of pipeline has been a regular bottleneck in both the oil and gas industries for a number of years.[104] There has also been a considerable cost escalation in pipeline construction in West Siberia. In the period 1965-1974 development of the pipeline system accounted for some two thirds of capital investment directed to the gas industry in this area.[105] In European operations the corresponding figure is one third.[106]

In order to alleviate the characteristic lag in technology and transport capacity, the Soviet Union has negotiated a number of agreements with Western companies. These have been directed to the solution of production problems and the construction of trunk pipelines. It is intended that credits used for the procurement of Western technology and equipment should be repaid in deliveries of oil and gas.[107] (The impact of existing and possible deliveries from the Soviet Union is discussed in Chapter 6.) Whatever the outcome of current negotiations, it is unlikely that large-scale development can be initiated and executed at a rate that will make a significant impact on the internal availability of oil and gas or on the export balance during the tenth Plan.[108]

There is little information in the tenth Plan on refining policy. The directives for the Plan indicated that primary refining capacity should rise by 25 to 30 per cent, and that the refinery mix should favour the output of high octane gasoline, low-sulphur distillate fuels, aviation kerosene and aromatics. This policy necessitates a substantial increase in secondary refining capacity.[109] It appears that the capital requirement for providing secondary refining capacity is two to three times more than for equivalent primary capacity. To minimize the increased capital requirement it is planned to expand existing secondary refining facilities rather than undertake the construction of new units. In response to the directive to site refining capacity close to consumption centers the capacity in the Ukraine is to be doubled, there is to be a 50 per cent increase in Belorussia, 80 per cent in Turkmenia and a fivefold increase in Kazakhstan.[110] The directives on the refinery mix for the tenth Plan were broadly similar to those set for the ninth, and reflect the trend towards maximum production of light, non-substitutable products at the expense of fuel oil, given the tightening supply of crude oil. Another significant factor discernable from the scarce information available, is that the trend towards increased capacity of individual processing units will continue, especially that of catalytic and steam crackers, the basis for manufacture of the much-desired light products.[111]

Oil and Gas Pricing, Its Relationship to Trends in the Energy Balance

Rising production costs, incurred as the oil and gas industries become increasingly dependent on operations in West Siberia, have highlighted the problem of the pricing of oil, refined products and natural gas in relation to other fuels. The question of crude oil pricing has been discussed in a recent article in 'Planovoe khozyaistvo'.[112] Current wholesale prices do not reflect fully the cost of production and a number of producers are therefore performing below the financial targets of 'khozraschet'. On average, only 60 per cent of costs incurred at the exploration and development stages are estimated to be covered by the present price structure. This results in the industry's inability to retain a level of surplus that could be recirculated as investment funds. It appears that this problem arose during the last two years of the ninth Plan.[113] The nature of the problem is, that, under the terms of the 1967 price reform, crude oil prices are determined on the basis of a calorific comparison with coal.[114] However, oil has an alternative value as a raw material that bears no relation to its value as a fuel. Therefore, the authors argued, the sole product for which the price should be determined in relation to coal is fuel oil.[115]

Currently the fuel oil element of the refinery balance is 30 to 35 per cent. Consequently, the authors argued, some 65 to 70 per cent of refinery output is being undervalued.[116] It could also be pointed out that as the proportion of light crude oils in the refinery balance increases, a substantial economic loss occurs in valuing the whole complement of refinery output on a calorific basis. An additional point is that under the present system of prices and returns it is economic to exploit only high productivity wells, leaving many unworked and thus further aggravating the tight supply position.[117]

At a time when the Soviet Union was seeking the increased use of fuel oil as boiler, furnace and power station fuel, its price was set at a level more attractive than that of coal, except in a number of the major coal-producing regions, where consumption of local coal was favoured.[118] Robert Campbell advanced the highly probable hypothesis that in the light of increasing difficulties in the oil industry the pricing policy 'may have been reversed' in favour of coal.[119] Given the general directives on fuel consumption policy already outlined, this would be a rational policy change on the Soviet part.

It appears that similar problems are about to be encountered in the gas industry. There are several gasfields where production is falling and some smaller deposits which have not been brought into production because of high basic costs. It is argued that the pricing system does not permit the desired level of financial efficiency at the enterprise level.[120] This too has been a function of recently rising development and production costs, bearing in mind that as late as 1974 the system as a whole was said to be recording a satisfactory financial performance.[121] However, given the rate at which consumption has grown and production has been accelerated in Siberia, the pattern of cost and return has changed very rapidly. Consequently it is argued that a more frequent reappraisal of price relativities is needed.[122] Though not explicit, many of the problems surrounding the valuation of petroleum products as feedstock rather than as fuel, affect the issue of natural gas pricing, since gas can fulfil either function. In any event higher prices for petroleum products and gas, relative to coal, would have the desired effect of improving fuel conversion efficiency and securing a move back towards coal in non-critical processes.

Rising world prices for energy, especially for oil, have altered the relative opportunity costs of production and export of energy raw materials in the Soviet Union. It will be seen from the analysis of the general trend in Soviet energy policy, given in Chapter 3, that the process of substitution of fuel oil by coal and natural gas was in progress prior to the OPEC price rises of 1973/1974. This substitution was influenced by the view that a gradual increase in world energy prices, especially of oil and gas, could be expected, though not on the scale that has occurred. The OPEC price rises served to compound this process.

However, given the slow rate of internal adjustment that is characteristic of the Soviet system, it appears that the domestic pricing structure has not yet altered to correspond with the desired direction for change, in response to the OPEC price rises. The production response to the price rises was that of seeking to conserve energy, particularly oil, and of enhancing the relative position of coal, natural gas and nuclear power.

The Eastern European Oil and Gas Industries 1976-1980

Eastern and Western analysts broadly agree that the prospects for increasing production of oil and gas in Eastern Europe are very limited. However, a measure of flexibility exists in the energy balance considered as a whole, though each member-country has different levels of indigenous reserves and forward energy requirements. These requirements depend on factors such as the share of output of energy-intensive industry within the total industrial output, and the relative efficiency in energy conversion.

Following the price rises imposed by the Soviet Union and OPEC, the energy plans of the Eastern European countries all emphasized the maximum development of indigenous reserves up to 1980. The Bulgarian Plan called for a 32 to 35 per cent rise in coal production between 1976 and 1980, by which time coal would account for approximately 20 per cent of the country's energy demand.[123] However, demand for oil and gas as a fuel or as chemical feedstock showed a twelvefold rise between 1960 and 1975, and the expansion of the chemical and petrochemical industries is expected to continue beyond 1980, requiring proportionally increasing amounts of oil and gas.[124] Despite exploratory drilling in the Black Sea there still appears to be no alternative to the expansion of imports in order to meet the rising demand for hydrocarbons. The preferred policy is that of joint development with fellow-members of Comecon.[125]

While the trend towards greater consumption of oil and gas in Hungary is likely to continue, it is now expected to do so at a lower rate than was anticipated before the oil price rises.[126] Consumption of natural gas is expected to reach 10 billion cubic metres in 1980, of which 6.0 billion will be produced domestically compared with 5.2 billion out of a total availability of 5.98 billion in 1975. It is intended to produce 2 million tonnes of oil per year to 1980, whereas coal production is expected to decline slightly to between 23.5-24.5 million tonnes in 1980 compared with 24.9 million in 1975. The net effect is that of raising import dependency in energy from 46 per cent in 1975 to 58 per cent in 1980.[127]

In meeting its growth targets for 1976-1980 the GDR sought to improve energy conversion efficiency by some 5 per cent annually.[128] Oil production will remain at its 1975 level, unless the joint Soviet-GDR-Polish enterprise currently exploring in the Baltic Sea discovers a field that can be brought into production rapidly.[129] Domestic production of natural gas is expected to expand to 14 billion cubic metres by 1980 and the target for production of brown coal of 250-254 million tonnes by 1980 represents a return to the production level attained in 1971.[130]

More detailed information has been provided by the remaining Eastern European countries who possess more extensive and relatively more varied reserves than the aforementioned countries.. Poland's reserves of hard coal can support a large expansion of production. The Plan for 1976-1980 envisaged substantial economic growth, within which industrial output was to rise by 48 to 50 per cent. However, the rate of growth in energy availability was intended to be lower than most other sectors of the economy, especially in the energy-intensive sectors.[131] The rapid development of indigenous coal and gas has become a priority in the Polish energy economy. In the case of hard coal, production in 1975 was 170 million tonnes[132] and this is expected to rise to 210 million in 1980.[133] Interest in developing brown coal reserves has revived and it is intended to raise output from 39.9 million tonnes in 1975, a figure which had remained roughly constant since 1972, to 48 million in 1980.[134] There has been a corresponding reorientation in hydrocarbon policy. In early 1973 the share of hard coal in the Polish energy balance was predicted to decline from 75 per cent in 1970 to 55 per cent by 1985 and 50 per cent by the year 2000, by which time oil and gas together would account for 30 per cent and nuclear power for 12 per cent of energy demands.[135] However, as a result of price and supply developments in Soviet and Middle East markets Polish planners revised their estimates to favour a lower (undisclosed) share to be held by oil in the period to 1980.[136] The forward position for Polish gas production is uncertain. There was little growth between 1971-1975 and little to suggest that there would be a marked increase in the 1975 production level of just under 6 billion cubic metres before 1980. However, the Plan target for 1980 was set at an ambitious level of 9 billion cubic metres. Growth in gas consumption will also depend on supply from the Soviet Union.

There are signs of growing tension in the Romanian energy balance, caused in the main by stagnation in domestic oil production. At the outset of the 1976-1980 Plan it was envisaged that the share of domestic energy demand to be met by indigenous production would fall from 86 per cent in 1974 to 75 per cent in 1980.[137] At the same time the major change in the domestic consumption pattern would be the rise in demand for oil both as energy and feedstock for the chemical industry. This industry's share of total oil demand was expected to rise from 15 per cent in 1975 to 21 per cent in 1980.[138] Whereas the average annual rate of growth in energy consumption during 1971-1975 was 7 per cent, the corresponding figure for 1976-1980 was put at 6.8 per cent, and the necessity of maximum development and utilization of domestic resources was stressed.[139] This latter element was reflected in individual fuel production targets for 1980. Oil production was scheduled to reach 15.5 million tonnes, showing little difference from the 1975 level. Likewise, natural gas production in 1980 was put at 26.8 billion cubic metres, only slightly above the 1975 level. (A further 7-9 billion cubic metres of wellhead gas can be expected.) However, the target for coal production in 1980 was put at 56 million tonnes, the substantial bulk of which is to consist of brown coal, an increase of some 75 per cent on 1975 levels.[140] This policy for primary energy production was complemented by targets for fuel conservation, particularly in the (energy-intensive) extractive and engineering industries.[141]

An additional effect of the deteriorating domestic supply position has been that of intensifying interest in nuclear power. Decisions concerning the commencement of power station construction taken during the drafting of the 1976-1980 Plan will not, however, alleviate the problem until the mid-eighties. Early estimates looked to a contribution by nuclear power to total electricity generation of at least 20 per cent by 1980.[142] This indicates Romania's considerable long-term commitment to this form of energy. At the time of writing there is some difficulty in assessing the effect of the major earthquake of 4 March 1977 on the short- and medium-term energy prospects. Information on the nature and extent of damage to installations or on the geological effect on individual fields has been sparse, but one authoritative Western source has expressed the opinion that damage was relatively slight and that it is unlikely to have any significant long-term effect.[143] It may be the case that the damage sustained by industry as a whole will result in demand for energy falling below expected levels, and that the resulting energy balance will not differ in terms of the relative share of fuels from that which was originally anticipated. In order to repair earthquake damage the Romanian government has allocated an additional 6,500 million lei to the oil industry in the remainder of the current Plan for reconstruction and also for new exploration, which is to include the Black Sea shelf.

Substantial changes in energy policy in Czechoslovakia are expected between 1976 and 1980. From 1971 to 1975 the entire increase in energy requirements was met by imports but the Plan for 1976-1980 called for greater domestic production of fuels. One third of the increase in energy consumption between 1976 and 1980 was to be provided by domestic resources, especially brown coal for use in electricity generation. Consumption of fuel oil in power stations was to be retarded and it is now intended that eventually coal will be the sole fuel used for new electricity generating capacity. Likewise in general boiler and furnace use, coal will be the preferred fuel even when gas is available in increased quantities after 1978, when the pipeline from the Orenburg field is commissioned.[144] Production of hard coal is relatively insignificant and its output is not likely to be raised between 1976 and 1980. Brown coal production was scheduled to rise from 86 million tonnes in 1975 to 95-100 million in 1980, operations in the Most area accounting for the major part of this increase. Czechoslovakia is dependent on the USSR for supplies of crude oil and natural gas, and demand for these will continue to grow. In the case of oil a significant change in the pattern of demand for refined products, complementing the relative decline in fuel oil, is the increased demand for naphtha to support the planned growth of the chemical and petrochemical industries. The share of oil consumption accounted for in these industries is to rise from 6 per cent in 1975 to 13 per cent in 1980.[145] In an attempt to alleviate the oil supply position Czechoslovakia has agreed to participate in the construction of the 'Adria' pipeline, which will deliver some 5 million tonnes of Kuwaiti oil per year after 1980.[146] It was estimated by the Czechs that on the results of exploratory work carried out during the 1971-1975 Plan, domestic production of oil and gas could not be increased between 1976 and 1980.[147] Beyond 1980 the development of nuclear power is seen to be the optimum solution to energy provision.[148]

Oil and Gas in Comecon 1976-1980: An Overview

The major Plan targets for fuel production in Eastern Europe for 1980 are detailed in table 5.9. Changes in the supply position in Comecon and the economic strain imposed by price rises have caused a shift in the direction of energy planning in the bloc. Even when OPEC oil was available at some 10 dollars per tonne in the late sixties the foreign exchange cost was greater than the cost of using Soviet oil supply. Despite the rising prices of Soviet oil and gas to Eastern

Table 5.9 1980 Plan Targets for Major Primary Energy Sources in Eastern Europe

Country	*Hydroelectricity (billion kWh)*	*Coal (million tonnes)*		*Oil (million tonnes)*	*Natural Gas (billion cubic metres)*
		Hard	*Brown*		
Bulgaria	3.5	–	38.0	0.5	1.0
Czechoslovakia	4.0	30.0	95.0	0.5	2.4
GDR	1.2	–	254.0	0.2	14.0
Hungary	0.2	3.0	21.0	2.0	6.0
Poland	2.5	210.0	48.0	0.4	9.0
Romania	10.0	7.0	49.0	15.5	35.5

Sources: Russell, *op cit*, pp 32, 118; Plan texts, except where output was too low for inclusion. The Plan text for Romania put the 1980 target for natural gas at 26.8 billion cubic metres, but this figure excluded wellhead gas, an estimate for which is included here. The target growth of Polish brown coal production (not cited in the Plan text) was assumed to be the same as for hard coal (22 per cent).

Europe there remains the fundamental benefit of a continued high level of supply. Given the considerable incentive present in the Soviet Union's continuing supply to Western markets, and the fact that the 'Druzhba' oil pipeline is operating at design capacity, the Soviet preference for providing gas rather than oil to Eastern Europe is the major element of change between 1976 and 1980. Prior to the commissioning of joint projects on Soviet territory and the further development of nuclear power, the hydrocarbon supply position in Comecon, whilst not of crisis proportions, will remain tightly balanced.

References

1. Marianna Slocum, 'Soviet Energy: An Internal Assessment', *Technology Review*, October/November 1974, p 17.

2. *ibid*, p 24.

3. *ibid*, p 29.

4. *ibid*, pp 29, 32.

5. P Hanson, 'The Soviet Energy Balance', *Nature*, 6 May 1976, pp 3-5.

6. *ibid*, p 3.

7. *ibid*, pp 3-4.

8. *ibid*, p 5.

9. Emily E Jack, J Richard Lee, Harold H Lent, 'Outlook for Soviet Energy', in *The Soviet Economy in a New Perspective*, Washington DC, US Congress, Joint Economic Committee, 1976, pp 460-478.

10. *ibid*, pp 460-461.

11. *ibid*, p 472.

12. *ibid*, p 473.

13. Central Intelligence Agency of the USA, *Prospects for Soviet Oil Production*, Washington, CIA, April 1977, (9 pp). See also *Prospects for Soviet Oil Production: A Supplemental Analysis*, Washington, CIA, July 1977.

14. CIA (April 1977), p 9.

15. *ibid.*

16. A A Makarov, L A Melent'ev, *Metody issledovaniya i optimizatsii energeticheskogo khozyaistva*, Novosibirsk Nauka, 1973.

17. *ibid*, pp 30-31.

18. P Neporozhnii, 'Perspektivy Sovetskoi energetiki', *Planovoe khozyaistvo*, 8/1975, pp 42-49. Neporozhnii stresses that at the outset of the ninth Plan, generation of kWh of electricity needed an input of 366 grammes of standard fuel. The target for 1974 was 345 grammes, which was achieved. An input of 341 grammes was anticipated by the end of 1975.

19. A Vigdorchik et al, 'Metody optimizatsii dolgosrochnogo razvitiya toplivno-energeticheskogo kompleksa SSSR', *Planovoe khozyaistvo*, 2/1975, pp 29-37.

20. *ibid*, p 37.

21. V A Shelest indicated that oil and gas production costs will eventually be two to four times and coal six to eight times lower in Siberia than in European Russia. *Regional'nye, energo-ekonomicheskie problemy SSSR*, Moscow, Nauka, 1975, p 205.

22. Makarov, Melent'ev, *op cit*, p 24.

23. Development and plans for Soviet nuclear power generation are outlined by A Albonetti, 'La Situation Enérgetique et Nucléaire en Union Soviétique', in *Round Table on the Exploitation of Siberia's Natural Resources*, Brussels, NATO, 1974, pp 113-120. See also P R Pryde, *Nuclear Energy Development in the Soviet Union*, paper presented to a Conference on Soviet and East European energy problems, University of Alberta, May 1977, 13 pp. Mimeo.

24. *Pravda*, 14 December 1975, p 2.

25. *Izvestiya*, 28 October 1976, p 3.

26. *Pravda*, 14 December 1975, p 2.

27. *ibid*, p 6.

28. *Pravda*, 18 December 1975, p 3, also *Ekonomicheskaya gazeta*, 22/1976, p 4.

29. *Ekonomicheskaya gazeta*, 22/1976, p 4.

30. *ibid.*

31. I Ya Vainer, 'Neftyanaya promyshlennost' v desyatoi pyatiletke', *Neftyanik*, 5/1976, p 2.

32. Z Ibragimova, 'Tyumenskii kompleks i ego budushchee', *Ekonomika i organizatsiya promyshlennogo proizvodstva*, 5/1976, p 8.

33. *ibid*, p 9.

34. *ibid*, p 20.

35. R I Kuzovatkin, 'Front i tyl Samotlora', *Ekonomika i organizatsiya promyshlennogo proizvodstva*, 6/1976, pp 78, 80.

36. *ibid*, p 81.

37. *Pravda*, 11 June 1975, p 3.

38. *Ekonomicheskaya gazeta*, 22/1976, p 4.

39. *Sotsialisticheskaya industriya*, 13 September 1974, p 2.

40. *Vyshka*, 16 January 1976, p 2.

41. *Sotsialisticheskaya industriya*, 29 July 1976, p 2.

42. N A Mal'tsev, 'Itogi i zadachi', *Neftyanoe khozyaistvo*, 3/1977, p 5.

43. *Pravda*, 12 June 1975, p 3. A recent Soviet paper on the development of gas condensate stated that reserves are sufficient to permit a 'sharp increase' in production in the short term. The authors stressed that use of condensate in preference to crude oil derivatives for petrochemical manufacture will 'simplify and cheapen' existing productive processes. Yu P Korotaev, G P Gurevich, I A Leont'ev, 'Dolgovremennoe obespechenie potrebitelei kondensatnym syr'em', *Gazovaya promyshlennost'*, 4/1977, p 21.

44. Note that *Neftyanoe khozyaistvo*, 2/1977, p 3, recorded that the 1976 oil production target was fulfilled at 100.3 per cent, showing 511.1 million tonnes against Plan of 509.4 million. The balance is the gas condensate element.

45. Mal'tsev, *op cit*, p 4.

46. *Ekonomicheskaya gazeta*, 8/1977, p 2.

47. *ibid.*

48. *ibid.*

49. A comprehensive description of the problems faced by the Soviet Union in developing offshore deposits of oil and gas was given by Joseph P Riva Jr in 'Soviet Offshore Oil and Gas', in *Soviet Oceans Development*, Washington DC, US Congress, National Ocean Policy Committee, 1976, pp 479-500.

50. Soviet Offshore output declined from 12.9 million tonnes in 1970 to 11.5 million in 1974. *ibid*, p 487.

51. *Pravda*, 14 December 1975, p 2.

52. *ibid*, p 6.

53. *Ekonomicheskaya gazeta*, 6/1977, p 1.

54. Editorial to *Gazovaya promyshlennost'*, 11/1976, p 2.

55. S A Orudzhev, 'Osnovnye zadachi razvitiya gazovoi promyshlennosti v 1976 godu – pervom godu desyatoi pyatiletki', *Gazovaya promyshlennost'*, 1/1976, p 3.

56. Jack, Lee, Lent, *op cit*, p 464.

57. Editorial, 'Po puti, namechennomu 25 s'ezdom KPSS', *Gazovaya promyshlennost'*, 4/1976, p 1.

58. Orudzhev, *op cit*, pp 5-6.

59. B P Orlov, 'Perspektivy Sibirskoi promyshlennosti', *Ekonomika i organizatsiya promyshlennogo proizvodstva*, 5/1976, p 42.

60. Alan B Smith, 'Soviet Dependence on Siberian Resource Development' in *The Soviet Economy in a New Perspective*, Washington DC, US Congress, Joint Economic Committee 1976, p 491.

61. *Gazovaya promyshlennost'*, 4/1976, p 1.

62. S F Gudkov, 'Nauchnye issledovaniya – dvizhushchaya sila teknicheskogo progressa', *Gazovaya promyshlennost'*, 1/1976, p 9.

63. Orudzhev, *op cit*, p 5.

64. N V Petlichenko, 'Dobycha gaza v pervom godu desyatoi pyatiletki', *Gazovaya promyshlennost'*, 2/1976, p 5.

65. P T Shmyglya, 'Nauka na sluzhbe osvoeniya gazovykh mestorozhdenii Sibiri', *Gazovaya promyshlennost'*, 2/1976, p 17.

66. G P Sulimenkov, D A Podluzskii, V S Bulatov, 'Problemy povysheniya effektivnosti osvoeniya Tyumenskikh mestorozhdenii gaza', *Gazovaya promyshlennost'*, 3/1976, p 12.

67. *ibid*, p 13.

68. For example, the forestry industry in Tyumen' was developed before oil and gas, and hence still has substantial delivery quotas to fulfil, directing wood to distant deficient areas. One result is that the oil and gas construction industry in Tyumen' faces a paradoxical shortage of wood, which is brought from other areas at very high cost. *ibid*, p 15.

69. Yu M Pavlov, 'Ekonomicheskie problemy razvitiya Sibiri i Dal'nego Vostoka', *Izvestiya AN SSSR, Ser. ekon.*, 2/1976, p 81.

70. Editorial, 'Osnovnye napravleniya razvitiya gazovoi promyshlennosti v iubileinom godu', *Gazovaya promyshlennost'*, 4/1977, pp 5-6.

71. S A Orudzhev, 'Osnovnye napravleniya povysheniya effektivnosti proizvodstva v otrasli v svete reshenii 25 s'ezda KPSS', *Gazovaya promyshlennost'*, 11/1976, p 3.

72. *ibid*, pp 3-4.

73. A A Annaliev, 'Turkmengazprom na novom etape razvitiya otrasli', *Gazovaya promyshlennost'*, 3/1976, p 10.

74. *Sotsialisticheskaya industriya*, 26 August 1976, p 1.

75. Annaliev, *op cit*, p 10; Orudzhev, 'Osnovnye zadachi . . .', *Gazovaya promyshlennost'*, 1/1976, p 6.

76. *Izvestiya*, 7 July 1976, p 1.

77. Editorial, 'Pretvoryaya v zhizn' plant Partii', *Stroitel' stvo truboprovodov*, 4/1976, p 3.

78. *Pravda Vostoka*, 23 December 1975, p 2.

79. Orudzhev, *Osnovnye zadachi . . .*, (1976), p 6.

80. *Stroitel'stvo truboprovodov*, 4/1976, p 2.

81. S A Orudzhev, 'Osnovnye zadachi razvitiya gazovoi promyshlennosti v iubileinom godu', *Gazovaya promyshlennost'*, 4/1977, p 6.

82. *Ekonomicheskaya gazeta*, 6/1977, p 2.

83. *Stroitel'stvo truboprovodov*, 4/1976, p 2.

84. G B Rassokhin et al, 'Opyt proektirovaniya i uskorennogo vvoda v razrabotku Vuktylskogo gazokondensatnogo mestorozhdeniya', *Gazovaya promyshlennost'*, 10/1976, p 31.

85. *Stroitel'stvo truboprovodov*, 4/1976, p 3.

86. Orudzhev, *Osnovnye zadachi . . .*, (1976), p 6.

87. Shelest, *op cit*, p 133.

88. Orudzhev, *Osnovnye zadachi . . .*, (1976), p 6.

89. *Pravda*, 28 January 1978, p 1.

90. *Stroitel'stvo truboprovodov*, 4/1976, p 1.

91. V G Chirskov, 'Problemy kruglogodlichnogo stroitel'stva truboprovodov v bolotistoi mestnosti', *Stroitel'stvo truboprovodov*, 1/1976, p 10.

92. V G Chirskov, 'Razvivaya truboprovodnyi transport Zapadnoi Sibiri', *Stroitel'stvo truboprovodov*, 4/1976, p 13.

93. *ibid.*

94. Editorial to *Stroitel'stvo truboprovodov*, 5/1976, p 3.

95. G I Gorechenkov, 'Vysokoe kachestvo proektirovaniya – vazhnoe uslovie rosta effektivnosti stroitel'stva', *Stroitel'stvo truboprovodov*, 5/1976, p 6.

96. Description of the agreement reached at the 28th Congress of Comecon, held in Sofia in June 1974, concerning joint participation in the construction of the pipeline is given in Appendix F of Jeremy Russell, *Energy as a Factor in Soviet Foreign Policy*, Farnborough, Saxon House, 1976, pp 229-231.

97. *Stroitel'stvo truboprovodov*, 1/1976, p 2.

98. *Hospodarske noviny*, 47/1975, p 8; also *Vneshnyaya torgovlya SSSR*, 8/1976, p 16.

99. J Bethkenhagen, 'Die Zusammenarbeit der RGW-Länder auf dem Energiesektor', *Osteuropa Wirtschaft*, 2/1977, p 71.

100. *Stroitel'stvo truboprovodov*, 4/1976, p 3.

101. *Stroitel'stvo truboprovodov*, 4/1977, p 1.

102. S S Ushakov, 'Tekhniko-ekonomicheskie problemy transporta topliva', *Moscow Transport*, 1972, p 15.

103. *ibid*, p 166.

104. Western sources have cited this as a factor contributing to the underfulfilment of Plan, especially in the gas industry. See Russell, *op cit*, pp 48, 66-67, and R W Campbell, *Trends in the Soviet Oil and Gas Industry*, Baltimore, Johns Hopkins Press, 1976, pp 36-37.

105. V S Bulatov, 'K voprosu ob udorozhanii gazoprovodnogo stroitel'stva na Tyumenskom severe', *Izvestiya SO AN SSSR, Ser. obshch. nauk.*, 1/1976, p 47.

106. *ibid*, p 49.

107. *ibid*, pp 49-50.

108. The state of negotiations between the Soviet Union and Western companies and governments was outlined in J D Park, 'Oil and Gas in Comecon', in E de Keyser (ed), *The European Offshore Oil and Gas Yearbook 1976/1977*, London, Kogan Page, 1976, pp 261-262. See also Appendix A.

109. *Izvestiya*, 14 December 1975, p 2.

110. *Ekonomicheskaya gazeta*, 13/1976, p 4. No data on actual capacities are divulged.

111. V B Yastremskaya et al, *Organizatsiya i planirovanie proizvodstva v neftyanoi i gazovoi promyshlennosti*, Moscow, Nedra, 1975, p 34.

112. S Levin, V Vasil'ev, N Kosinov, 'Tsenoobrazovanie v neftyanoi promyshlennosti', *Planovoe khozyaistvo*, 7/1976, pp 110-115.

113. *ibid*, p 111.

114. This was discussed in R W Campbell, *Trends in the Soviet Oil and Gas Industry*, Baltimore, Johns Hopkins Press, 1976, pp 189-192.

115. Levin, Vasil'ev, Kosinov, *op cit*, pp 111-112.

116. V I Nazarov, A A Il'inskii, N I Pimenov, 'K obosnovaniyu predel'nykh ekonomicheskikh pokazatelei poiskov i razvedki neftyanykh plastov', *Ekonomika neftyanoi promyshlennosti*, 4/1977, p 3.

117. *ibid*, p 7.

118. V I Torbin, *Territorial'naya differentsiyatsiya tsen v tyazheloi promyshlennosti*, Moscow, Ekonomika, 1974, p 175.

119. Campbell, *op cit*, p 70. Note also that the price of gasoline has been increased substantially in the Soviet Union since 1975 partly to offset rising costs in the industry, partly to induce conservation.

120. I S Tyshlyar, A A Kevorkov, 'Ispol'zovanie mekhanizma dogovornykh tsen na neft' dlya bolee polnoi otrabotki mestorozhdenii', *Gazovaya promyshlennost'*, 3/1977, p 23.

121. As analyzed by Campbell, *op cit*, pp 70-71.

122. Tyshlyar, Kevorkov, *op cit*, p 24.

123. I Kondov et al, 'Razshiryavaneto na energiinata baza i rastezhot na neinata efektivnost', *Planovo stopanstvo*, 2/1976, p 37.

124. *ibid*, p 38. See also Cecil Rajana, *The Chemical and Petrochemical Industries of Russia and Eastern Europe 1960-1980*, London, Sussex University Press, 1975, p 179.

125. Kondov et al, *op cit*, p 38.

126. V Zurbuchen, 'Die fünfte Fünfjahrplan der Ungarischen Volksrepublik 1976 bis 1980', *Wirtschaftswissenschaft*, 10/1976, p 1324.

127. *ibid.*

128. *Petroleum Economist*, April 1976, p 137.

129. *Słowo powszechne*, 27 November 1975, p 2, records the setting-up of this consortium.

130. G Schurer, 'Hauptrichtungen der weiteren Vervollkommnung unserer materiell-technischen Basis in den Jahren 1976 bis 1980', *Einheit*, 9/1976, p 1015.

131. J Kopytowski, Polish Deputy Minister of the Chemical Industry in an interview with 'Petroleum Economist', *Petroleum Economist*, January 1976, p 19.

132. *ibid.*

133. *ibid.*

134. *Polish Economic Survey*, 7/1975, p 6.

135. J Mitrega, 'Die Brennstoff- und Energiebasis Polens heute und im Jahre 2000', *Montan-Rundschau*, 3/1973, p 72.

136. *Petroleum Economist*, January 1976, p 19.

137. I V Herescu, 'Creşterea economică şi consumul de energie', *Revista economică*, 34/1975, p 12.

138. I V Herescu, 'Dezvoltarea bazei energetice (1)', *Revista economică*, 28/1976, p 1.

139. *ibid*, pp 1-2.

140. (i) *Scînteia*, 3 August 1974, p 2; (ii) *Petroleum Economist*, January 1975, p 30; (iii) *Scînteia*, 3 July 1976, pp 2-4.

141. I V Herescu, 'Dezvoltarea bazei energetice (2)', *Revista economică*, 29/1976, p 1. See also V Popescu, 'Economisirea şi valorificarea resurselor naturale', *Revista economică*, 7/1976, p 5.

142. E Rodean, 'Energia nucleară şi promovarea progresului tehnico-ştiinţific în ţara noastra', *Era socialistă*, 3/1975, p 40.

143. *Petroleum Economist*, April 1977, p 156.

144. *Hospodářské noviny*, 10 September 1976, p 3.

145. *ibid*. Also *Rudé právo*, 26 May 1976, p 5, and Rajana, *op cit*, p 179.

146. Russell, *op cit*, pp 228-229.

147. *Preliminary Observations on Long-Term Problems in the Field of Basic Products and Energy and of Long-Term Trends in the Czechoslovak Economy*, document transmitted by the Government of Czechoslovakia to the Economic Commission for Europe, Ref. EC.AD (XI)/R.4/Add. 12, 30 January 1974, p 2.

148. *ibid*, also V Ehrenberger, 'Perspektivy rozvoje palivové a energeticke základny', *Nová Mysl*, 7/8/1976, pp 100-101.

Chapter 6

Soviet Trade in Oil and Gas, 1970-1980

In Comecon as a whole the level of foreign trade dependence is low by world standards, and the Soviet Union is considerably less dependent on foreign trade in relation to net material product than fellow members of Comecon. In the sixties the rate of growth in foreign trade of the Comecon countries was below that of Western industrialized countries, and also below the world average rate. Since 1972 the position has been reversed. Whilst world trade expanded in 1972 by some 18 per cent, Comecon imports rose by 22 per cent and exports by 17.9 per cent. The member countries enjoyed more favourable prices in hard currency trade due to advantageous realignments in Comecon currencies in relation to sterling and the US dollar.[1]

A major source of influence determining the recent development pattern of Soviet foreign trade is the low level, by world standards, of many branches of indigenous technology. As far as capital equipment is concerned, this 'technology gap' prompted the import of Western goods and expertise and in the early sixties, as this process gathered pace, there was limited scope for balancing the drain on the economy through exports.

In the consumer goods sector of the Soviet economy the low level of indigenous technology provides finished goods of limited export potential. There is also evidence of rising demand for consumer goods, including those of Western origin. In order to alleviate problems in the balance of trade and payments the Soviet Union could direct a higher proportion of indigenous resources to the development of industrial capital goods and to the greater financing of research. However, there is doubt as to the Soviet Union's ability to do this on the scale required since the demands on resources posed by the development of the military sector, and the undesirable social consequences that might arise as a result of excessive restraint on consumer demand, appear to be equally influential considerations in the overall balance.

One possibility open to the Soviet Union is that of initiating joint-venture projects with fellow members of Comecon in the industrial sector as a whole. An alternative might be to raise loans from elsewhere in Comecon, but this is limited by the fact that these countries are themselves in difficulties in respect of the balance of trade and payments. They are likewise aware of the dangers of attempting to restrain consumer demand. The possibility of increasing exports over a range of goods is fraught with difficulties: capital goods need to be retained for domestic use, consumer goods tend to be uncompetitive in world markets and agricultural exports are unlikely to expand in view of the periodic and unpredictable shortfalls encountered in this sector. Raising further finance through Western capital markets is limited in scope, since the Comecon currencies remain only partially convertible and the level of indebtedness of some countries is already regarded as high. Though the Soviet Union has relied for the most part on Western credits negotiated on a state-to-state basis, more recently loans have been raised on the Eurodollar market, since the US government has been cautious about extending credits. On the other hand Western European governments and institutions have extended substantial credit facilities. At a time of Western economic slump credits are more readily obtainable by Comecon members, at a time of buoyancy more readily repayable.

Exports of raw materials, especially fuels, are a different proposition. The rapid rise in prices for energy raw materials has enabled the Soviet Union, in the short-term, to recoup much of the cost of financing the import of technology and foodstuffs. In 1975, the Soviet Union exported some 130 million tonnes of crude oil and refined products out of a total crude oil production of 491 million. Of this 63 million tonnes were exported to the Eastern European full members of Comecon and 67 million to the remainder of the world. Over the period from 1971 to 1975 the share of fuels and energy in total Soviet export value rose from 17.9 to 31.4 per cent.

The issues affecting the export balance in the period to 1980, when oil production is planned to reach 640 million tonnes, are the rates of increase in domestic consumption, the demand for energy in Comecon, the degree of flexibility enjoyed by the Soviet Union in meeting the increase in demand, and the capacity of the Eastern European countries to supplement deliveries from the Soviet Union with purchases from OPEC. Such considerations will be influenced by the Soviet Union's analysis of its need for hard currencies and the extent to which this can be met by expanding direct exports to the corresponding markets and by enhancing its role as an oil and gas broker.

Perspectives on Soviet Oil and Gas Trade

As early as 1966 the Polish analyst S Albinowski advanced the view that the Soviet Union would run into an oil supply crisis by 1980.[2] He based his calculations on a projected rise in per capita consumption of oil from 870 kilogrammes in 1965 to 2 tonnes in 1980, at which time he estimated the Soviet population would have risen to 280 million. Corresponding figures for Eastern Europe in 1980 were 1.5 tonnes with a population of 115 million. Albinowski concluded that demand for oil in the Soviet Union would reach 560 million tonnes, and in Eastern Europe 170 million in 1980. He put Soviet production in 1980 at 630 million tonnes and that of Eastern Europe in the same year at 33 million. Assuming that the Soviet Union would allocate some 20 million tonnes for export to non-Comecon markets, leaving 50 million tonnes for Eastern Europe, an oil deficit of 90 million tonnes is predicted in the latter area in 1980.[3]

Albinowski's extrapolation of per capita consumption of oil to 1980 overlooked not only the potential for improvement in energy conversion efficiency in general, and particularly in the use of oil, but also the likelihood of a declining rate of population growth. More important is the fact that although at the time of Albinowski's estimate the Soviet gas industry was in its infancy, the rates of growth recorded had inspired confidence. An increasing share in the energy balance was forecast for this fuel. In addition the Soviet foreign trade monopoly had entered into preliminary negotiations on gas export contracts.

A somewhat different analysis of Soviet trade potential was published in 1971 by the West German analyst Werner Gumpel.[4] He pointed to the wide range of options which existed at the time of his writing, for the alleviation of supply tensions in any of the Soviet Union's fuels industries through coordinated planning and investment. Gumpel acknowledged an increasing tension in oil supply and demand, and in the range of opinion as to the future of Soviet oil trade. He noted that an (unattested) OECD forecast put 1980 oil production at 690-710 million tonnes and demand at 613 million, leaving a surplus of 75 to 95 million tonnes for export.[5] He pointed out that a prominent Soviet energy economist N V Mel'nikov had estimated in 1969 that Soviet production in 1980 would be no more than 607 million tonnes,[6] and that a TASS bulletin of 16 May 1969 referred to projected Soviet oil production in 1980 as reaching 'more than 500 million tonnes'.[7] Gumpel favoured acceptance of the Mel'nikov forecast, suggesting a small domestic oil deficit of 5 or 6 million tonnes in 1980. Though Gumpel subscribed to the view that the Soviet Union would be a net importer of crude oil by 1980,[8] the area of discussion being the extent of dependence, he did make reference to the possible emergence of natural gas, and stressed the more favourable ratio of reserves to production and the increasing export potential.[9] He did not, however, offer any analysis of domestic and export substitution possibilities, seeing this as a minor issue.[10] A major point to which Gumpel alluded, but left undeveloped, was that the Soviet Union might be prepared to incur higher domestic costs in the energy sector in order to remain independent, as far as possible, of supply from an area as historically unstable as the Middle East.[11] Albinowski had pointed out, in 1966, that greater involvement on the part of Comecon countries in the Middle East, primarily in hydrocarbon production, might afford the opportunity of influencing the development of the latter towards socialist-type patterns, as a counter to the spread of Western colonialism.[12] Albinowski confined his view of the Soviet Union's possible import of Middle East oil to suggesting that the issue might become part of a global competition for geopolitical influence: Gumpel, however, intimated a possible extension of the 'Brezhnev doctrine' to include a direct Soviet military presence in the area.[13]

The analysis of Robert E Ebel, completed in 1970, presented yet one more view. Though primarily concerned with developments in Soviet oil and gas trade from 1917 until 1968, the work did offer, in conclusion, an estimate of Soviet oil and gas trade potential. Ebel was the only author writing in this period to stress the impact of the development of gas on the production of, and demand for, oil and petroleum products. He pointed to the low prime cost of producing natural gas compared with other fuels[14] and to the potential of gas to take the place of oil as a first choice substitute for coal, in accordance with Soviet policy in the late sixties. He also stressed its importance as an export fuel in its own right in Comecon and Western markets. The rapid growth of the gas industry would, he argued, depress the rate of growth in demand for oil in the bloc.[15]

Ebel estimated the figure for Soviet oil production in 1975 at 460 million tonnes, with the expectation that this would rise to 600-620 million by 1980.[16] Despite this acceptance of a lower figure for 1975 than his contemporaries were putting forward (a figure which turned out to be 45 million tonnes less than the Plan target eventually set for 1975), Ebel saw no basis for the proposition that the Soviet Union would be obliged to import substantial quantities of oil from the Middle East in the period up to 1975. He suggested rather that increased imports might be contemplated for economic reasons only and on the basis of accepting oil as the repayment element in barter trade.[17] A study of Ebel's forward estimate of gas production and trade sheds some light on his view of the prospects for the oil industry. The inaccuracy of Ebel's projections were due to some extent to the difficulties faced at that time by Soviet planners themselves. In a short space of time the production level projected for 1975 was revised from a range of 380-400 billion cubic metres[18] to 300-340 billion[19] and to 280-300 billion.[20] The last estimate proved accurate. Equally important were the Soviet estimates made at that time of production in 1980. A projection made in 1961 forecast a likely level of 680-700 billion cubic metres:[21] this was scaled down to 640-650 billion in an estimate made at the end of 1967.[22] These estimates proved to be well in excess of the final objective, which limited the rate of increase of the share of natural gas in the energy balance and in exports. Ebel forecast a slow-down in sales of oil to non-Comecon markets,[23] counterbalanced by rising exports of gas, especially to the expanding markets of the Netherlands, France, West Germany and Belgium, which were supplied at that time from the Dutch Groningen gasfield.[24] The range of Ebel's estimate of 7-9 billion cubic metres as the net Soviet export surplus of gas in 1975 proved accurate.[25]

An analysis carried out in the early seventies by a Vienna-based team stressed the export of raw materials as having been advantageous to Comecon as a whole, and especially so for the Soviet Union. However, the growth in output appeared inadequate to sustain the rate of economic growth in the industrial sector thought likely at the time. Consequently, the authors argued, trade in such materials, especially in coal, oil and gas, could be expected to decline.[26] They pointed to a declining volume of net exports of oil and refined products from the Soviet Union caused by increasing imports from the Middle East[27] and stressed that while the Soviet Union could benefit from the rise in world prices for raw materials, the opportunities presented by the rise in oil prices in particular, might not be fully exploitable.[28]

However, analysis based on historical trends overlooks the changing role of oil and gas exports in relation to the demands of the economy as a whole. Their role as the export commodity most likely to secure the means of acquiring much-needed technology is the subject of a later review article by the Oxford economist Michael Kaser.[29] His view was that oil and gas form the 'cornerstone' of Soviet plans for the development of trade with the West, to the extent that the quantity of each available for export to Eastern Europe would be only a residual quantity after consideration of domestic needs and the level of imports from the West necessary to support the economic Plan.[30] He argued that as long as the Soviet Union could maintain steady growth in oil production the volume of oil and refined products available for export to Eastern Europe should not decline below the 1970 level.[31] The factors determining the ultimate level of exports of oil to Western Europe, however, might be non-economic and related to the extent to which the West might be prepared to become dependent on the Soviet Union for the supply of so vital a material as oil. A determining factor would be the fear of a possible embargo by the Soviet Union and its effects.[32]

The greater involvement of the Soviet Union and Eastern Europe in world trade, and specifically in oil, in the changed financial market following the massive inflow of wealth and purchasing power to OPEC members after 1973, is the subject of a further article by the same author.[33] Kaser advanced the view that OPEC dollar surpluses might be deposited in Comecon institutions, offering financial stability and institutional solvency guaranteed by the state mechanism, and an unblemished record on the part of the Soviet Union in dealings with foreign creditors and depositors. Such a course of action on the part of OPEC would, he argued, allow OPEC to acquire goods without investment in Western institutions, a process which would serve merely to shore up the economic system of perceived political adversaries.[34] This perspective overlooked the possibility that OPEC, or rather many of its Arab members, might be equally wary of financing the furtherance of Soviet imperialism, given the considerable antipathy towards Soviet-type socialism existing in many quarters of the Arab world. Direct investment by Middle Eastern countries in Comecon has been limited and the latter group has had in fact to compete for a share of the petro-dollar surplus via the international banking system.

A study completed in 1975 provided a particularly detailed analysis of Soviet trade in energy raw materials, including substantial data on oil and gas.[35] The work covered the period from 1960 to 1980, but the most recent data on which it is based were for mid-1973. Bethkenhagen's projection of the supply pattern rested on the assumption that the world price of oil could be expected to rise to eight dollars per barrel (in 1973 dollars) by 1980.[36] This analysis contrasted with that of the majority of observers of the time: he maintained that in 1980 Eastern Europe would receive 100 million tonnes of oil and refined products from the Soviet Union, with an upper limit of 35 million tonnes from other sources.[37] Even with an assumed forward price for OPEC oil below that actually imposed, Bethkenhagen stressed the limited capacity of Eastern Europe to expand trade with OPEC producers. His projections were dependent on a number of (self-admitted) assumptions that simplified the forward position. For example he assumed that there would be no import restrictions on the part of Western Europe, and no provision of technology and know-how from the West that could affect Soviet export capacity before 1980.[38] He argued that Romania would remain independent of Soviet supply through to 1980[39] and that the upper limit on Soviet imports from the Middle East would be 20 million tonnes.[40] He pointed to the substantial export potential of gas in both Eastern and Western European markets, estimating that on the basis of contracts concluded up to the end of October 1973, exports of Soviet gas to Western Europe (West Germany, France, Italy, Finland, Austria and Sweden) would total 21 billion cubic metres in 1980.[41] His short-term estimate for oil production and exports proved to be somewhat inaccurate. He underestimated Soviet production in 1975, at 475 million tonnes, and total exports at 101 million, with the supposition that there would be no increase in exports to the West on the estimated figure for 1975 of 43.6 million tonnes.[42] The short-term forecast for gas was more accurate, though again he underestimated 1975 production slightly, putting this at 284 billion cubic metres, and overestimated imports and domestic demand. He took the view that imports of gas in 1980 would show little difference on his estimated level for 1975, with expansion in export trade from an estimated 22 billion cubic metres in 1975 to 50 billion in 1980.[43]

A similar perspective was presented at about the same time as the Bethkenhagen analysis by the American analyst A W Wright.[44] His point of departure was that Soviet decisions concerning the development of an export policy for oil and gas were determined by rational economic criteria. Whilst admitting an appreciation on the part of Soviet planners of potential 'political' gains resulting from trade policy (not only energy trade policy), he noted the existence of a point at which the incurrence of economic loss was judged excessive in relation to perceived political gain.[45] He argued that members of any 'customs union' may, in respect of certain individual products, pay a price for membership quantifiable in terms of import and export opportunities foregone, but that the overall benefit of membership may outweigh this.[46] He did, however, admit that in the course of the ninth Plan the terms of trade in energy materials altered in the Soviet Union's favour, correctly anticipating the Soviet Union's imposition of a substantial price rise for fuels supplied to Eastern Europe.[47] Wright made two important points. He expressed the view that the Soviet Union was capable of implementing fuel production and import/export policies involving independent development of indigenous resources, but that such decisions were required to be economically sound in relation to domestic cost, and to import and export prices of fuels. He maintained that the resurgence of interest on the part of the Soviet Union in joint ventures with Western associations was prompted by uncertainty as to whether the high import prices for energy in the post-1973 period could be expected to fall. This would leave the Soviet Union with a substantial commitment to (uneconomic) energy developments, whereas joint investment would both spread the risk and create a communality of interest in keeping world energy prices high. The second point was that Soviet trade in non-Comecon energy markets had been conducted on a sound commercial basis, with no evidence of price-cutting beyond the level required to secure a contract, and that this policy was likely to continue.[48]

Peter Odell considered the principal feature of the Soviet energy economy as a continuing drive for self-sufficiency, to the extent of incurring expense in supplying certain areas with indigenously produced fuel in preference to lower cost imports. He also maintained that the Soviet Union chose to import oil and gas for reasons of logistic convenience rather than of actual need.[49] This latter point appears justified in the case of gas. However, in the case of oil the position is more complicated. Whereas Soviet foreign trade statistics show imports of crude oil and refined products from certain Middle East and North African producers, it has been pointed out that some of this may in fact be delivered to Comecon or non-Comecon markets on Soviet account.[50]

Soviet Trade in Oil and Gas 1971-1975

The structure of imports and exports is no less regulated a phenomenon than domestic production within the Plan. The Soviet economy requires that the volume and composition of imports and exports be determined in relation to the requirements of domestic Plan fulfilment. Unlike the domestic Plan, however, the Soviet Union releases no data on the import-export Plan. It is known that import-export agreements are usually timed to coincide with the Plan period and that prices are fixed for the duration of an agreement. It is also important to bear in mind, as J Wilczynski has pointed out,[51] that although the export price of a particular Soviet commodity may be lower than its domestic counterpart, the Soviet Union does not sustain economic loss, since the crucial factor is the net surplus to the national production account. The opportunity cost concept is, however, increasingly followed by Soviet planners. The price rises imposed by OPEC in 1973 and subsequently, changed substantially the Soviet Union's previous patterns of long-term energy trade and price relationships with her customers and suppliers.

The role of fuels and energy within the commodity structure of Soviet foreign trade during the ninth Plan is outlined in table 6.1. It will be seen that not only have fuels become the major product group in Soviet export trade but they have also constituted the biggest growth sector in value terms. In import trade the measure of the Soviet Union's self-sufficiency and flexibility in energy planning is illustrated by the maintenance of the percentage value of fuels imports in relation to total imports, resulting in a much-decreased volume, after the OPEC oil price rises of 1973 and 1974. The enhanced earning power of the export of energy raw materials is evident from the export trends.

A major consideration of Western analysts during the ninth Plan has been the gradual increase of Soviet and Eastern European indebtedness to hard currency trading areas. Given the prevalent view in Western circles that Comecon was facing a growing energy deficit, especially in oil, the financial capability of the bloc to sustain imports of oil became an issue of growing interest. The possibility of further expansion in trade with the West is hampered by quota and tariff restrictions. There are difficulties associated with the export of Soviet-agricultural products and foodstuffs due to the inadequacy of the sector itself to meet domestic demand and to problems associated with the European Common Agricultural Policy. A recent estimate of likely developments in trading patterns made by Richard Portes concluded that if Soviet imports remained at the level of 1976 the Soviet Union would need to increase exports to hard currency markets at the rate of 10 per cent per year in order to level off the debt on hard currency accounts by 1980.[52]

The essential attractiveness of raw material exports from the Western standpoint is that the products themselves are homogenous, and meet simple, readily testable quality standards in contrast to Soviet technological goods, the quality and performance of which are often below world standards. There are two incentives for Western countries to become more involved in negotiations to secure the supply of Soviet oil, refined products and gas: first, that the changing role of OPEC within the world petroleum market has given rise to uncertainty

Table 6.1 Commodity Structure of Soviet Foreign Trade 1971-1975 (%)

(A) Exports

	1971	*1972*	*1973*	*1974*	*1975*
Machinery & Equipment	21.8	23.6	21.8	19.2	18.7
Fuel	17.9	17.7	19.2	25.4	31.4
Ores & Concentrate Metals	18.7	19.0	17.1	14.7	14.3
Chemicals, Fertilizers	3.4	3.3	3.0	3.6	3.5
Wood Products, Cellulose	6.3	6.1	6.4	6.9	5.7
Textiles, raw and semi-finished	3.3	3.8	3.3	3.3	2.9
Furs	0.4	0.4	0.3	0.3	0.2
Food Products	9.2	5.9	5.6	7.1	4.8
Consumer Goods	2.9	3.1	3.0	2.9	3.1
Total	83.9	82.9	79.9	83.4	84.6
Unspecified	16.1	17.1	20.3	16.6	15.4
Total in million rubles	12,429	12,734	15,802	20,738	24,030
Of which Socialist countries	8,116	8,286	9,115	11,092	14,584

(B) Imports

	1971	*1972*	*1973*	*1974*	*1975*
Machinery & Equipment	34.0	34.6	34.3	32.4	33.9
Fuel and Energy	2.7	3.0	3.4	3.5	4.0
Ores, Concentrates, Metals	9.8	8.9	9.9	13.6	11.5
Chemicals & Fertilizers	5.4	4.9	4.3	6.3	4.7
Wood Pulp, Cellulose, Paper	2.1	1.8	1.6	1.9	2.2
Textiles, raw and semi-finished	4.5	3.3	3.7	4.1	2.4
Food Products	15.2	18.0	20.2	17.1	23.0
Consumer Goods	20.1	18.6	15.9	14.6	13.0
Total	93.8	93.1	93.3	93.5	94.7
Unspecified	6.2	6.9	6.7	6.5	5.3
Total in million rubles	11,232	13,303	15,544	18,834	26,669
Of which Socialist countries	7,360	8,519	9,216	10,304	13,986

Source: *Narodnoe khozyaistvo SSSR*, corresponding years

over the future of oil supply and its price, and second, that the availability of Soviet gas in Western Europe might be the most economical way of supplying the established market, given that production from the Dutch Groningen gasfield is expected to peak and decline before 1980.

It has been pointed out that the bilateralism that characterizes Soviet foreign trade has no basis in ideology or theory, but rather in its relevance to a country lacking reserves of hard currency which desires to remain relatively isolated from the vacillations of world markets. In addition it facilitates the integration of trade into the planning process.[53] The latter point is expanded in a study by Adam Zwass, in which he suggested that the policy of using foreign trade solely to fill gaps in domestic supply of essential imports was conceived when the Soviet Union considered capitalism a short-term threat that would be destroyed through its own inherent contradictions.[54] Nove argued further that the major reason for the continuation of this policy is that the central Plan is still the basis of economic activity, into which foreign trade must be fitted in order that planners may demonstrate the formulation of an economic strategy that is not subject to world market forces.[55]

The issue of pricing is particularly pertinent to Soviet decisions concerning the pattern of exports of oil and refined products. The supply of these became more restricted in relation to demand during the seventies than they were in the sixties. And it is considered that gas export opportunities may have been lost through delays in the construction of delivery facilities. Zwass pointed out that production for the home and export markets was regarded as a single function. Enterprises manufacture under exactly the same terms and criteria for export as for the home market, without taking into account economic conditions prevailing in their particular export market.[56] In fact the Soviet foreign trade system shields the enterprise from world market fluctuations by absorbing any short-term gains or losses into the Budget.[57]

The prime objective in intra-Comecon trade is that of balancing a number of production relations in physical terms, but which, for the purpose of internal accounting, are expressed in monetary values. Zwass gave some details of the mechanism used in determining prices in intra-Comecon trade.[58] He stated that in addition to the fact that prices bore no relation to domestic cost and relative scarcity, the underlying problem was the potential obsolescence, during the time of their application, of the prices used,[59] plus a system of exchange rates that did not reflect the international competitiveness of the economy of the individual country.[60] The case of oil pricing has been particularly interesting. During the late sixties there were a number of monopolistic elements determining the 'world price' of oil in that major oil companies operating in the Middle East worked increasingly with OPEC to secure an orderly rise in the price of oil. How the Soviet Union defines a 'monopolistic element' to be discounted in calculating the 'world price' is uncertain: what is, however, pertinent is that the rapid change in the market price of oil after 1973 prompted the Soviet Union to take advantage of OPEC's monopolistic success in that the value of Soviet energy exports was reappraised. This enabled the Soviet Union to take advantage of opportunities in hard currency trade and to revalue its increasingly labour-intensive oil and gas against goods supplied in return by other members of Comecon.

However, there was some uncertainty regarding the Soviet Union's capacity to maintain the rate of growth of production and trade that had been recorded in the sixties. At that time the quadrupling of prices for oil and refined products had greatly enhanced opportunities for increasing earnings of hard currencies. The question then arose whether the Soviet Union would forego such opportunities in order to supply the increasing requirement of Comecon, or alternatively (and additionally), whether Soviet oil and gas would alleviate the impact of higher prices on developing countries.

The pattern of Soviet trade in oil and refined products is detailed in table 6.2, the major features being the steady growth of gross and net exports, the rising trend of deliveries to Comecon and non-EEC markets, and the fluctuating but apparently non-expanding level of trade with the EEC nine. The OPEC price rises of 1973 and 1974 are seen to have had a marked effect in halting the growth of Soviet imports from the Middle East. As internal energy policy was readjusted to take these price rises into account, the upward trend was resumed in 1975 but at a lower level than suggested by the trend of the preceding few years.

Table 6.3 details the corresponding pattern of Soviet trade in natural gas, showing the gradual transition in the status of the Soviet Union from net importer to net exporter, favouring primarily the Comecon market for logistic as well as politico-economic reasons. The significance of Soviet imports of gas is twofold. First, deliveries of Iranian gas constitute repayment in kind for Soviet supply of equipment and expertise used to develop the Iranian gas industry. Soviet engineers supervised the construction of pipeline from Iran that joins the Soviet system linking Central Asia with the Central and European areas. The gas is consumed in Armenia and Georgia, where local production is declining and where no substantial new reserves have been discovered recently. Similarly gas from Afghanistan serves as repayment for a variety of Soviet machinery and technology, and is consumed in the industrial areas of Kirghizia and Tadzhikistan,

Table 6.2a Soviet Trade in Oil and Oil Products 1971-1975 (million tonnes)

Exports	*1971*	*1972*	*1973*	*1974*	*1975*
Comecon	44.76	48.89	55.28	58.71	63.28
EEC 9	24.52	23.67	26.63	20.96	25.03
Rest of World	35.82	34.44	36.69	36.53	42.02
Total	105.10	107.00	118.30	116.20	130.35
of which: crude oil	74.80	76.20	85.30	80.60	93.07
refined products	30.30	30.80	33.20	35.60	37.28
Total Imports	6.70	9.10	14.70	5.40	7.56
of which: crude oil	5.10	7.80	13.20	4.40	6.50
refined products	1.50	1.30	1.50	1.00	1.06
Net Exports	98.40	97.90	103.60	110.80	122.79

Source: *Vneshnyaya torgovlya SSSR*, corresponding years

Table 6.2b Exports of Oil and Oil Products by Country of Destination 1971-1975
(Volume in million tonnes, unit prices in rubles per tonne)

Destination	1971		1972		1973		1974		1975	
	Volume	*Unit Price*	*Volume*	*Unit Price*	*Volume*	*Unit Price*	*Volume*	*Unit Price*	*Volume*	*Unit Price*
(i) Comecon										
Bulgaria	7.96	14.6	7.95	14.9	9.32	14.5	10.86	15.1	11.55	34.2
GDR	10.38	13.6	11.48	14.1	12.99	14.1	14.42	21.1	14.95	28.2
Poland	9.55	16.6	11.06	16.5	12.34	17.4	11.86	20.7	13.27	39.5
Czechoslovakia	11.81	16.2	12.87	16.4	14.34	16.3	14.84	16.3	15.97	30.9
Hungary	5.06	16.6	5.53	17.0	6.29	17.8	6.73	21.0	7.54	41.0
(ii) EEC 9										
France	4.54	15.1	3.08	14.1	5.35	17.0	1.36	61.9	3.31	58.2
West Germany	6.09	15.4	6.20	14.4	5.85	36.6	6.34	65.7	7.63	62.2
Netherlands	1.63	17.7	2.43	16.5	3.22	42.5	2.98	71.4	3.09	65.0
Belgium	2.04	15.1	2.52	14.2	1.67	41.0	1.75	66.4	1.26	63.1
Italy	9.00	13.1	8.43	12.9	8.65	17.4	6.79	58.2	6.88	57.0
Great Britian	0.03	26.5	0.05	15.4	0.83	20.7	0.92	76.0	1.50	64.7
Irish Republic	0.33	9.7	0.19	10.6	0.18	26.8	0.12	62.5	0.18	49.0
Denmark	0.86	12.0	0.77	13.2	0.63	51.5	0.70	61.6	1.18	57.6
(iii) Rest of World										
Austria	1.13	15.9	0.97	15.4	1.25	22.8	0.97	63.2	1.33	58.7
Afghanistan	0.15	37.6	0.16	33.4	0.17	29.0	0.19	31.0	0.15	86.6
N Vietnam	0.38	34.2	0.19	45.0	0.23	42.4	0.29	39.0	0.40	35.3
Ghana	0.60	14.5	0.63	14.1	0.61	14.7	0.31	78.5	0.14	65.6
Guinea	0.07	29.5	0.07	28.8	0.09	28.7	0.08	82.8	0.06	75.3
Greece	1.01	16.6	0.91	18.9	0.80	22.0	1.03	67.0	1.89	59.5
Egypt	1.60	19.8	1.44	18.2	0.35	27.8	0.23	77.0	0.23	82.2
India	0.47	20.8	0.38	18.7	0.48	22.5	1.01	65.9	1.21	77.5
Iceland	0.38	22.7	0.44	19.9	0.47	27.0	0.46	75.2	0.45	73.5

Table 6.2b continued

Destination	1971		1972		1973		1974		1975	
	Volume	*Unit Price*	*Volume*	*Unit Price*	*Volume*	*Unit Price*	*Volume*	*Unit Price*	*Volume*	*Unit Price*
Spain	0.21	18.9	0.78	14.6	0.51	17.8	1.35	60.6	1.72	57.6
Cyprus	0.20	13.1	0.13	14.0	0.12	14.2	0.11	52.2	0.21	44.9
N Korea	0.70	33.4	0.40	34.6	0.59	28.6	0.94	26.4	1.11	24.0
Cuba	6.44	11.4	7.02	13.1	7.44	15.2	7.64	17.7	8.06	30.8
Morocco	0.87	13.9	0.93	14.3	0.94	14.4	0.65	69.8	0.65	58.9
Mongolia	0.27	42.1	0.30	36.6	0.32	37.8	0.35	35.9	0.36	36.1
Norway	0.63	18.3	0.45	15.1	0.60	19.2	0.28	74.8	0.28	63.5
Somalia	0.07	23.4	0.07	28.5	0.07	25.7	0.11	38.6	0.12	72.6
Turkey	0.07	28.7	neg	–	–	–	–	–	–	–
Finland	8.57	19.4	8.63	18.8	10.03	22.1	9.17	66.8	8.77	61.9
Switzerland	0.80	20.8	0.82	17.0	0.66	55.3	0.78	67.6	0.96	64.2
Sweden	4.57	12.9	4.36	13.3	3.22	18.0	3.03	55.9	3.45	50.1
Yugoslavia	2.88	18.2	3.40	15.1	3.89	22.8	3.80	65.5	4.44	61.3
Japan	3.28	13.6	1.01	15.4	2.02*	27.9	1.24*	57.8	1.32	51.2
West Berlin	0.17	15.6	0.36	16.0	0.43	52.5	0.53	76.3	0.85	70.7
Bangladesh	–	–	0.03	17.0	0.05	20.5	0.17	70.1	0.17	77.5
Nigeria	0.15	15.8	neg	–	–	–	–	–	–	–
USA	–	–	–	–	–	–	0.18	59.8	0.54	62.2
Portugal	–	–	–	–	–	–	0.07		1.06	59.0
Nepal	–	–	–	–	–	–	neg	–	0.06	72.5
Syria	neg	–	neg	–	neg	–	0.05	76.7	neg	–
Liberia	–	–	–	–	neg	–	neg	–	0.03	51.5
Brazil	–	–	–	–	–	–	1.23	65.0	1.48	88.7
Canada	–	–	–	–	–	–	0.16	50.8	0.22	60.2

**plus 1.0 million tonnes in 1973 and 0.2 million in 1974 under a non-commercial agreement*
neg = negligible

Source: *Vneshnyaya torgovlya SSSR*, corresponding years

Table 6.2c Pattern of Soviet Trade in Crude Oil and Major Refined Products 1971-1975 (million tonnes, unit prices in rubles per tonne)

	1971		*1972*		*1973*		*1974*		*1975*	
	Volume	*Unit Price*	*Volume*	*Unit Price*	*Volume*	*Unit Price*	*Volume*	*Unit Price*	*Volume*	*Unit Price*
(i) Exports										
Crude Oil	74.8	14.1	76.2	14.4	85.3	15.8	80.6	28.9	93.1	40.3
Gasoline	4.1	22.1	4.5	19.8	5.5	28.9	5.8	68.4	6.0	67.1
Kerosene	2.4	24.4	2.3	22.5	2.3	33.5	2.6	55.3	2.6	67.6
Gas Oil	11.4	22.4	12.5	20.2	14.2	44.9	15.8	64.3	15.9	66.8
Fuel Oil	11.9	12.6	11.1	12.6	10.4	14.1	10.8	38.9	12.0	38.1
(ii) Imports										
Crude Oil	5.1	11.1	7.8	14.3	13.2	16.7	4.4	66.1	6.5	61.1
Gasoline	0.37	28.5	0.34	27.0	0.53	24.8	0.48	26.7	0.48	53.0
Kerosene	0.33	24.7	0.17	29.0	0.17	29.0	0.17	29.0	0.18	59.0
Gas Oil	0.18	24.4	0.14	25.2	0.14	25.4	0.12	25.2	0.12	51.0
Fuel Oil	–	–	–	–	–	–	–	–	–	–

Source: *Vneshnyaya torgovlya SSSR*, corresponding years

where local fuel production, especially of hydrocarbons, is inadequate to meet local demand. In the case of Iranian and Afghan supply the Soviet Union has successfully negotiated to maintain a lower import price than is obtainable for Soviet exported gas. Second, there is no alternative market for either Iranian or Afghan gas. The major gas-consuming centers, namely Western Europe and the US, have gas available at prices well below those at which Iranian or Afghan gas could be supplied. The bulk of Iranian gas is associated with oil and the alternative would be to flare it at the wellhead.

The Soviet Union committed itself to supply fellow members of Comecon between 1971 and 1975 with 243 million tonnes of oil and 33 billion cubic metres of gas, compared with 138 million tonnes and 8 billion cubic metres over the previous Plan.[61] It will be seen from tables 6.2 and 6.3 that deliveries of oil and refined products to Comecon between 1971 and 1975 totalled 271 million tonnes and deliveries of gas 30.5 billion cubic metres. Since the export statistics published by the Soviet Union and most of the East European countries no longer record separate deliveries of oil and oil products by country, it is not possible to calculate the extent of divergence from the agreed schedule. However, given that refinery capacity in Eastern Europe is regarded as adequate to meet domestic demand, with a consequently minor need for refined products, the likelihood is that the substantial part of deliveries consisted of crude oil and that contractual obligations were broadly honoured in spite of the increasing difficulties in oil production and the shortfall against the original Plan. Bearing in mind the causes of the shortfall in the Soviet gas industry's production, the delivery of 30.5 billion cubic metres to other Comecon countries during the ninth Plan represented a considerable commitment.

An interesting feature of the Soviet Union's pattern of trade in petroleum products is the maintenance of similar ratios between the product groups. Unit values for Soviet export trade in oil and refined products reflected the price adjustments made in non-Comecon trade in 1974 following the OPEC price rises, and in trade with Comecon from 1 January 1975. The impact of the OPEC price rises was reflected in unit values for imported oil in 1974 and 1975 and in the

Table 6.3a Soviet Trade in Natural Gas 1971-1975
(billion cubic metres)

	1971	*1972*	*1973*	*1974*	*1975*
Exports					
Comecon	3.13	3.44	4.07	8.56	11.29
EEC 9	–	–	0.35	2.93	5.44
Rest of World	1.43	1.63	1.62	2.55	2.60
Total Exports	4.56	5.07	6.83	14.04	19.33
Imports	8.18	11.04	11.41	11.94	12.41
Net Exports	*(–3.62)*	*(–5.97)*	*(–4.58)*	*2.10*	*6.92*

Source: *Vneshnyaya torgovlya SSSR*, corresponding years. (Note that in 1973 the totals of the individual countries' exports do not sum to the reported national total. The discrepancy of 740 million cubic metres is not explained in Soviet statistics.)

Table 6.3b Gas trade by Country 1971-1975
(billion cubic metres, unit prices in rubles per thousand cubic metres)

	1971		1972		1973		1974		1975	
	Volume	*Unit Price*	*Volume*	*Unit Price*	*Volume*	*Unit Price*	*Volume*	*Unit Price*	*Volume*	*Unit Price*
Exports										
(i) Comecon										
Bulgaria	–	–	–	–	–	–	0.31	15.16	1.19	29.39
Hungary	–	–	–	–	–	–	–	–	0.60	
GDR	–	–	–	–	–	–	2.90	15.10	3.30	15.11
Poland	1.49	13.84	1.50	13.84	1.71	13.84	2.12	13.86	2.51	28.01
Czechoslovakia	1.64	13.81	1.94	14.75	2.36	14.75	3.23	15.28	3.69	25.47
(ii) EEC 9										
Italy	–	–	–	–	–	–	0.79	7.81	2.34	16.35
West Germany	–	–	–	–	0.35	14.62	2.14	14.12	3.10	17.83
(iii) Rest of World										
Austria	1.43	12.69	1.63	11.60	1.62	10.46	2.11	13.62	1.88	30.33
Finland	–	–	–	–	–	–	0.44	47.73	0.72	46.85
(Note that in 1973 the individual quantities do not sum to the reported global total. The shortfall of 760 million cubic metres is not explained in Soviet foreign trade statistics.)										
Imports										
Afghanistan	2.51	5.43	2.85	4.99	2.73	4.94	2.85	6.84	2.85	12.43
Iran	5.62	6.19	8.20	6.29	8.68	7.80	9.04	14.57	9.56	15.33

Source: *Vneshnyaya torgovlya SSSR*, corresponding years

drop in imports from the 1973 level. The relatively small volume of refined products imported by the Soviet Union originated in Romania and the unit prices in 1975 reflected the Romanian response to the re-evaluation within intra-Comecon trade of oil and its derivatives.

The terms of the change in the method of determining oil prices, effective from 1 January 1975, were that prices would be fixed *annually* on the basis of average world prices for the preceding five years. An exception was made in 1975, when it was decided that prices would be determined against the average of the three-year period from 1972 to 1974.[62] Though information on the effect of the Soviet oil price rise has been somewhat fragmentary, it has proved possible to make a number of observations on its impact on petroleum consumption in Eastern Europe and on Soviet trade. Prior to the Soviet price rise, events had been set in motion that were to anticipate the further effect of such a rise. The impact of the OPEC rises brought about an immediate program of energy conservation in Eastern Europe. In view of the small percentage of oil and energy supply accounted for by OPEC oil at that time, it might be reasonably supposed that the reaction related as much to an anticipated increase in these deliveries as to the solution of the immediate problem. For example, in Czechoslovakia the prices of the range of refined products were doubled, and previously allocated delivery levels from refineries were cut by an average of 10 per cent.[63] In Romania, where the domestic supply of energy, and of oil in particular, was a good deal less tight than elsewhere in Eastern Europe, strict norms for improvements in energy conversion efficiency were introduced, and these were expected to remain in force for the 1976-1980 Plan.[64]

The 1975 price for deliveries of Soviet oil to Eastern Europe was set at some 130 per cent above that charged from 1971 to 1974. The benefits of this substantial price rise to the Soviet Union were the earlier availability of higher revenues, and the catalyzing effect on discussions being held in Eastern Europe concerning the forward development of the energy balance of the individual countries plus the issue of fuel conversion efficiency in general. The price rise involved an increase from an average of 16 rubles per tonne at the Soviet border to 37 rubles, with slight differences between the countries in order to take account of the varying transportation costs.[65] There was a further problem in addition to the price rise, namely that for supply of oil above the negotiated annual quotas for 1975 and subsequently, the Soviet Union required payment in hard currencies.[66]

The price rise for Soviet oil was imposed at a time when Eastern Europe was faced with a mounting trade and payments deficit with the West, and its trade surplus with the Soviet Union was moving towards a deficit. The deficit with the West was incurred when the Eastern European countries adapted a development strategy of import-led growth. This was done in differing degrees by each East European country and in order that output of imported machinery could eventually be used to repay the credits granted for its procurement. Decisions were taken in the late sixties to initiate this process. However, the upheavals in raw material prices between 1973 and 1975 placed unanticipated stress on the international payment capacity of the Eastern European countries.

The process of import-led growth was developed to the greatest extent by

Poland. Fortunately, the existence of substantial reserves of hard coal enabled planners to react to the increases in oil prices by reorienting energy policy to favour a slower rate of substitution of coal by hydrocarbon fuels. The other Eastern European countries, with the possible continued exception of Romania, will need to allocate a level of goods for export to the Soviet Union proportionally higher than hitherto. They will also need to seek credits from the Soviet Union to offset the increased fuel import bill if Soviet absorption capacity proves inadequate. This will be necessary since a continued rise in Soviet oil prices can be expected between 1976 and 1980 following the introduction of the moving average as the basis for calculation of prices. This is offset, albeit to a limited degree, in that one effect of the OPEC price rises has been that of exacerbating the detectable decline in the economic health of the West, to the extent that decreased industrial demand, including demand for machinery and technological processes, has prompted greater efforts on the part of the West to sell these items to Eastern Europe. The West is prepared to do so on credit terms that are now substantially more favourable than could have been anticipated in the late sixties and early seventies, when a steady growth was normal in the West.

The Soviet Union reacted quickly to the OPEC price rises of 1973-1974. As an immediate measure imports of oil were cut from 13.2 million tonnes in 1973 to 4.4 million in 1974. At the same time the Soviet Union incurred a cut in exports of oil and refined products from 118.3 million tonnes in 1973 to 116.2 million in 1974. Complex issues, such as the likely level of the future cohesion of OPEC, its effect on pricing policy and the extent to which high world prices for energy could be expected to hold, were debated within the context of the options open to the Soviet Union for alternative energy policies. These debates included the planned change in the role of oil in relation to gas and coal.

The availability of Soviet natural gas in Eastern Europe dates from the latter part of the eighth Plan, when contracts were negotiated with Poland and Czechoslovakia. By the end of the ninth Plan Soviet gas was available to each of the member countries of the bloc with the exception of Romania, which remained self-sufficient up to 1975, and even exported a small quantity to Hungary. Comparison of unit import and export prices shows a diverging trend in favour of the Soviet Union, though as world prices hardened in the aftermath of the oil price rises, both Afghanistan and Iran succeeded in negotiating higher prices for their gas delivered to the Soviet Union.[67] Given the more favourable ratio of explored reserves to production of gas compared with oil, and the lower percentage of exports in production, a policy of substitution of oil by gas, not only in domestic but also in export markets, was initiated. The underfulfilment of targets by the gas industry during the ninth Plan delayed this process. However, the emergence of supply problems was probably instrumental in forging cooperation between the Comecon members in developing Soviet oil and gas resources, the results of which will be evident towards the end of the tenth Plan.

In contrast to the Eastern European energy markets, which showed falling rates of growth in consumption, the energy markets of the rest of the industrialized world experienced an actual decline in energy consumed, and this decline was most marked in the case of oil. Western Europe has been the prime market

Table 6.4 Impact of Soviet Oil and Oil Products on Principal West European Markets 1971-1975 (million tonnes)

Country	*1971*	*1972*	*1973*	*1974*	*1975*
France:					
Total Imports*	115.74	126.84	142.28	137.64	114.00
from USSR	4.54	3.08	5.35	1.36	3.31
West Germany:					
Total Imports	135.63	141.75	153.77	141.46	129.06
from USSR	6.09	6.20	5.85	6.34	7.63
Netherlands:					
Total Imports	69.02	67.79	72.16	64.59	55.23
from USSR	1.63	2.43	3.22	2.98	3.09
Belgium:					
Total Imports	37.99	42.38	45.67	39.67	38.03
from USSR	2.04	2.52	1.67	1.75	1.23
Italy:					
Total Imports	120.24	124.39	123.63	126.25	104.84
from USSR	9.00	8.43	8.65	6.79	6.88
Great Britain:					
Total Imports	126.80	128.14	133.28	127.59	104.16
from USSR	0.03	0.05	0.83	0.92	1.50
Irish Republic:					
Total Imports	5.81	5.62	5.06	5.81	5.57
from USSR	0.33	0.19	0.18	0.12	0.18
Denmark:					
Total Imports	21.07	22.01	21.31	20.14	18.77
from USSR	0.86	0.77	0.63	0.70	1.18
Austria:					
Total Imports	7.39	8.07	9.22	8.36	8.56
from USSR	1.13	0.97	1.25	0.97	1.33
Spain:					
Total Imports	36.53	28.34	44.08	45.84	43.85
from USSR	0.21	0.78	0.51	1.35	1.72
Norway:					
Total Imports	8.94	9.92	10.65	9.70	8.69
from USSR	0.63	0.45	0.60	0.28	0.28
Finland:					
Total Imports	11.43	12.97	13.59	13.63	12.83
from USSR	8.57	8.63	10.03	9.17	8.77
Switzerland:					
Total Imports	13.02	13.43	14.38	13.35	12.67
from USSR	0.80	0.82	0.66	0.78	0.96
Sweden:					
Total Imports	29.75	28.66	28.64	27.39	30.88
from USSR	4.57	4.36	3.22	3.03	3.45

**Note that in each country imports include crude oil and refined products.*

Sources: *Table 6.2*; *Eurostat, Energy Statistics 1975*, pp 118-119, 163; *Eurostat, Quarterly Energy Statistics*, 4/1976, pp 51-55, 61-65; *OECD Statistics of Energy*, 1959-1973 passim, 1960-1974 passim, 1973-1975 passim

for Soviet oil since the Soviet Union re-emerged as an exporter in the late fifties. Western Europe was also seen as an obvious market for natural gas. In the immediate aftermath of the OPEC price rises and embargo on Rotterdam, the principal distribution point for the Western European oil market, the view was advanced that the Soviet Union might choose to expand sales substantially even at the expense of Eastern Europe.[68]

An analysis of the impact of Soviet oil on principal Western European markets is given in table 6.4. The essential features are a rise in the Soviet share of the total market, still small in most cases, as total demand declined (see table 6.2) and the marked increase in unit prices after 1973. The absolute volume of Soviet exports rose only slightly during this period, but included some fluctuations. However, the major change between 1971 and 1975 was the effect on the visible trade balance of the increased price obtainable for Soviet oil after 1973. Data illustrating the importance to the Soviet Union of trade in oil with Western European countries are presented in table 6.5.

Finland and Sweden are the only Western European countries in which Soviet oil and petroleum products can be said to have made a significant contribution to supply. In the case of Finland there has been a long history of trade with Tsarist Russia and the Soviet Union, facilitated by geographical proximity and in more recent times by closer inclusion in the Comecon trading system when granted the opportunity of accounting in transferable rubles.[69] Trade with Sweden, linked with the earliest barter agreement for large-diameter pipe, was maintained through the ninth Plan, but showed signs of decline towards the end of the Plan.

Table 6.5 Soviet Visible Trade Balance with Principal West European Countries 1971-1975 (million rubles)

Country	*1971*	*1972*	*1973*	*1974*	*1975*
France:					
Exports	194.3	194.0	272.2	397.9	495.7
Oil Exports*	68.3	43.4	91.0	84.2	192.6
Oil % of Exports*	35.2	22.4	33.4	21.2	38.8
Imports	281.9	350.3	449.4	543.1	800.8
West Germany:					
Exports	254.7	255.9	453.8	834.5	857.9
Oil Exports	94.1	88.9	214.1	423.3	475.3
Oil % of Exports	36.9	34.8	47.2	50.7	55.4
Imports	411.9	571.4	756.4	1,374.2	1,919.4
Netherlands:					
Exports	153.6	154.6	260.6	394.3	303.8
Oil Exports	28.9	40.2	135.6	202.0	201.0
Oil % of Exports	18.8	26.0	58.9	51.2	66.1
Imports	70.5	67.7	95.6	176.4	147.2
Belgium:					
Exports	97.9	108.1	194.3	297.9	243.6
Oil Exports	30.9	35.7	68.6	116.2	79.2
Oil % of Exports	31.5	33.1	35.3	39.0	32.5
Imports	59.8	66.4	160.0	305.5	286.2

Table 6.5 continued

Country	*1971*	*1972*	*1973*	*1974*	*1975*
Italy:					
Exports	233.1	228.0	309.5	597.6	638.0
Oil Exports	117.7	109.0	153.1	394.9	392.3
Oil % of Exports	50.5	47.8	49.5	66.1	61.5
Imports	261.5	235.5	304.1	539.6	788.0
Great Britain:					
Exports	407.7	371.1	540.6	690.5	591.1
Oil Exports	0.7	0.8	17.3	69.9	97.3
Oil % of Exports	0.2	0.2	3.2	10.1	16.5
Imports	200.0	186.7	174.6	199.6	368.2
Irish Republic:					
Exports	4.3	2.9	5.6	12.3	10.7
Oil Exports	3.3	2.0	4.8	7.5	8.6
Oil % of Exports	75.3	67.6	85.0	60.7	80.6
Imports	0.3	0.5	0.6	15.9	20.4
Denmark:					
Exports	26.0	24.9	53.8	78.1	105.7
Oil Exports	10.4	10.2	32.5	43.1	67.8
Oil % of Exports	39.8	40.4	60.4	55.2	64.2
Imports	21.9	24.0	26.1	41.8	40.4
Austria:					
Exports	90.9	82.7	99.6	166.0	218.5
Oil Exports	18.0	14.9	28.6	61.4	77.9
Oil % of Exports	19.7	18.0	28.7	37.0	35.6
Imports	81.4	80.8	84.7	173.6	226.6
Spain:					
Exports	10.0	20.9	30.0	122.0	143.6
Oil Exports	4.0	11.5	9.1	81.7	100.3
Oil % of Exports	40.4	54.8	30.3	67.0	69.8
Imports	9.0	22.9	10.6	27.4	45.4
Norway:					
Exports	42.1	21.8	34.7	46.8	64.4
Oil Exports	11.5	6.8	11.5	20.9	18.0
Oil % of Exports	27.3	31.1	33.1	44.7	27.9
Imports	16.0	16.1	16.5	45.1	65.9
Switzerland:					
Exports	35.9	30.7	67.9	79.5	89.5
Oil Exports	16.8	14.0	36.1	52.8	61.6
Oil % of Exports	46.7	45.6	53.1	66.4	68.9
Imports	74.7	90.5	99.8	165.1	240.8
Finland:					
Exports	322.8	297.6	415.1	937.6	918.2
Oil Exports	166.4	162.1	221.8	614.8	542.5
Oil % of Exports	51.6	54.5	53.4	65.6	59.1
Imports	246.3	304.1	362.3	602.1	837.3
Sweden:					
Exports	110.0	108.9	130.7	285.8	289.5
Oil Exports	59.1	58.0	58.3	168.0	172.9
Oil % of Exports	53.7	53.3	44.6	58.8	59.7
Imports	85.5	79.9	101.6	149.7	255.9

**In each country the category 'oil' includes crude oil and refined products.*

Source: *Vneshnyaya torgovlya SSSR*, corresponding years

Soviet trade statistics show that the Soviet Union did not use the OPEC production cutback and eventual embargo on deliveries to the Netherlands as an opportunity for substantially increasing its deliveries. The 1973 volume was less than 1 million tonnes above that of 1972. However, the Soviet Union did benefit from the price rise. As Nove has indicated,[70] over half the Soviet Union's expanded trade with the Netherlands in 1973, consisted of oil and oil products bought at a price some 150 per cent above the 1972 level. This suggests that the majority of the Dutch purchases were made after the price rises and the imposition of the embargo.

Facing the problems of rising costs in the domestic oil industry and an apparently growing need for imports from the West, the Soviet Union's policy on oil business in Western Europe has been that of reaping the rewards of rising prices and of gaining an increased share of a declining market. This could well be maintained if the market began again to expand. Western Europe might then view the Soviet Union as no less secure a source of supply than the Middle East. However, any attempt by the Soviet Union to increase substantially its exports to Western Europe would prompt OPEC to adjust its pricing and output policies in order to retain control of the Western European oil supply balance. It seems that the Soviet Union appreciated the commercial consequences of such action.

As a result of the policy of continued growth in deliveries to Eastern and Western European markets, the Soviet Union was unable to consider the possible option of alleviating the effect of the OPEC price rises on developing countries. However, the absolute volume of Soviet oil exports to developing countries is less significant than the periodic negotiations for possible supply which induce existing suppliers and sellers of oil and oil products, often the affiliates of the major companies, to lower their prices to these countries.[71] There is some measure of advantage in this practice (notably the supply to India in the early sixties outlined in Chaper 2). However, the Soviet desire to maintain the unity of Comecon and the need to secure the supply of Western technology appears now to have taken precedence over the possibility of competing for politico-economic influence in the developing world, at least as far as using indigenous oil as an element of strategy.

In view of the particularly severe impact of the events of 1973-1974 on Japan it might seem surprising that trade in oil and refined products did not rise towards the end of the ninth Plan, but in fact showed a decline. An explanation of this trend is to be found in the emergence of China as an oil exporter, with Japan as its largest market. Deliveries of Chinese crude oil to Japan commenced in 1973 and totalled 1 million tonnes in that year. This figure rose to 4 million tonnes in 1974 and to 8 million in 1975.[72] The Japanese, influenced perhaps by the difficulties they were facing in negotiating supply from the Soviet Union at that time, which involved changing commitments on the Soviet part as to quantities, price and terms of payment, sought and secured marginal supplies from China on a barter basis. These were mainly in return for steel and industrial products and did not involve negotiating long-term credits and joint ventures with the Chinese.

In the aftermath of America's 'oil crisis' there was some discussion on the possibility of the Soviet Union supplying oil to the US in return for grain.[73]

One interpretation of this option, advanced at the outset of the 1975 Soviet-American negotiations over grain supply, was that the US had no immediate need for Soviet oil, but that its increasing presence on the world market might serve to bring down the price set by OPEC. It was thought that this would contribute to Kissinger's 'grand design' to restore economic equilibrium in the world and to 'fashion a new balance of power'.[74] However, despite the apparent need and desire of the Soviet Union to acquire a supply of American grain there was no great eagerness to sell domestically produced oil and refined products without a corresponding commitment by America to enter into joint projects on Soviet territory. Such a commitment would provide for the long-term development of Soviet resources. A further obstacle was that the US attempted initially to negotiate for supplies at below world prices. Russell argued that the dissipation of American interest, both in short-term purchases of spot cargoes and in long-term cooperation, was due to their suspicion of Soviet motives following the 'grain robbery' of 1972. The Americans were also preoccupied at this time with the socio-political impact of Watergate and the enthusiastic launching of 'Project Independence 1980'.[75]

There is, however, a different dimension to Soviet interest in the American oil market. Although official Soviet trade statistics recorded exports of small quantities of oil and refined products to the US in 1974 and 1975 (see table 6.2), these figures cannot be regarded as representing the total sum of business contracted between the two countries. Any Soviet oil or refined product made available to the US would be phased into the world logistic system of the major oil companies and might be presented as having been sold to any one of a number of countries. Similarly any oil made available to the Soviet Union in return for technical assistance, or as the balance of a barter arrangement, could be delivered from a third country through an oil broker to the US on Soviet account. Although the Soviet system of import classification provides for the inclusion of materials re-exported without having entered the Soviet Union, it records only quantities authorized directly by the Soviet Foreign Trade Ministry. Any 'switched' oil might be ordered by the purchaser through the international broking network and thus recorded as a direct delivery. It is likely, though not statistically demonstrable, that quantities of oil made available to the Soviet Union were shipped from Brazil and Venezuela to the US under such arrangements. It is not possible therefore to deduce the extent of the short-term commitment of 'Soviet' oil to the US.

During the ninth Plan the status of the Soviet Union as a gas trader changed from that of net importer to net exporter. This change was based on the delivery systems constructed as elements of barter trade during the latter part of the eighth Plan. These systems facilitated supply to Czechoslovakia and Austria with an extension of the pipeline to Northern Italy, to Poland and to West Germany with a branch-line supplying the chemical complexes of the Halle-Leipzig area in East Germany.

After the oil price rises of 1973 and 1974, gas prices in Western Europe, the Soviet Union's prime market, began to harden. The attractiveness of gas as an export fuel was further enhanced by the fact that the prices for imports of Afghan and Iranian gas did not rise proportionally to export prices (see table 6.2).

Export pricing of natural gas shows some interesting differences. In deliveries to Eastern Europe prices remained at a level of 13 to 15 rubles per thousand cubic metres from 1971 through to 1974, followed by a substantial rise, parallel to that for oil, in 1975. This applied in each of the Eastern European countries with the exception of East Germany. The likely explanation of this is that it had been intended to supply East Germany at an earlier stage, but commencement was delayed because of the underfulfilment of the production Plan by the Soviet Union, leaving a lower quantity available for export. Hence when deliveries did commence, the previously negotiated price was charged. Deliveries to Western countries are tied to a previous supply of steel pipe and equipment and the Soviet Union may have decided therefore that longer-term interests would be better served in meeting contracted deliveries to the West.

In the case of deliveries to Austria and Finland pricing to the former showed a marked rise in 1975 compared with 1974, and in the latter a substantially higher price than for any other country. In respect of Austria the higher price was negotiated under a contract signed in 1974 following the expiry of the previous one negotiated in 1966/1967, which linked gas deliveries to the supply of steel pipe. The price for return deliveries of gas related to the market conditions of the time. Decisions taken in late 1974 in respect of deliveries for 1975 and beyond, involved a re-evaluation of gas against the Austrian market price for fuel oil. Subsequent Soviet gas prices reflected not only the effect of the OPEC oil price rises but also the fact that since Austria is landlocked, the delivered cost of hydrocarbon fuels is, as with Switzerland, the highest in Western Europe.[76] In the Finnish case, the high cost of Soviet gas is due to the fact that supply is not linked to the counter-delivery of pipeline or equipment and has been negotiated separately from general bilateral trade agreements. The price was therefore linked to what the nearest free market alternative would have been.[77] In view of the rising cost of OPEC oil, and the absence of convenient alternative fuels, Finland had to pay the price determined by the international market.

During the ninth Plan the development of the Soviet gas industry was such that despite the shortfall against original targets planners were able to forecast trends in the domestic energy balance and trade structure that provided for its increasing contribution. The growing impact of Soviet gas on the energy economies of the individual Eastern European countries could not alleviate the tensions caused by shortcomings in production and supply of other fuels, and the rising cost of oil imports. Nonetheless the Soviet Union has negotiated supply contracts in Eastern and Western Europe that show considerable expansion over 1975 levels.

The pattern of Romanian trade in petroleum products is outlined in table 6.6. Although crude oil production in Romania showed little growth between 1971 and 1975, in accordance with the Plan, decisions had been taken to expand refinery capacity to 25 million tonnes per year with the intention of increasing exports of fuels and petrochemical feedstock. Despite the imposition of economy measures in 1974 domestic consumption grew rapidly and this resulted in declining net exports at a time when prices in hard currency markets were rising. Consequently quantities of refined products that had been earmarked for export were consumed domestically and, as outlined in Chapter 5, the need to arrest

Table 6.6 Romanian Trade in Oil and Oil Products 1971-1975 (million tonnes)

	1971	*1972*	*1973*	*1974*	*1975*
Crude Oil Availability	15,651	17,101	18,330	18,924	19,675
Total Exports of Products	5,143	4,966	4,841	6,456	6,057
including:					
Gasoline	588	579	610	1,106	1,464
Gas Oil	2,620	2,374	2,195	2,378	2,024
Fuel Oil	1,541	1,615	1,715	2,656	2,258

Sources: *Table 4.21*; *Statisticheskii ezhegodnik stran-chlenov SEV*, 1976, p 389; *Annuarul statistic din R S România*, 1976, p 391

this trend influenced the planning of the Romanian energy balance for 1976-1980.

As the difficulties in maintaining indigenous production of oil appear to be increasing, it is likely that a greater quantity of oil, imported from the Middle East, will be directed to domestic consumption in the medium-term. This would take place prior to the delivery of Soviet gas after 1980 and the further contribution of nuclear power to the energy balance, scheduled for the mid-eighties.

Soviet Trade in Oil and Gas 1976-1980

There are considerable differences in the views of Western analysts as to the likely level of exports of Soviet oil and gas in 1980, and these are outlined in tables 6.7 and 6.8. The issues are further complicated by the analysis of the Hungarian energy economist Istvan Dobozi, published in January 1973.[78] He too estimated an oil and gas deficit in Comecon as a whole, by 1980, and the likelihood of a deficit in the Soviet Union itself. Dobozi's analysis is detailed in table 6.9.

Since the Soviet Union records no data on consumption of individual fuels and stock changes, the quantity of oil and gas available for consumption, that is production net of imports and exports, has been taken as the basis for analysis, and the trends for the period 1971 to mid-1977 are outlined in table 6.10.

Western estimates of Soviet energy consumption made in the early to mid-seventies tended to the view that the rate of growth in consumption, especially of oil and gas, would be greater than the rate of growth in production,[79] and that this would result in a decline in exports.[80] The actual trends ran counter to expectation, as shown in the data contained in table 6.10. There has been a substantial degree of substitution of oil by gas, despite the failure of the gas industry to fulfil the initial targets of the ninth Plan. The major effect, however, has been a lower level of total energy demand than anticipated, due to the declining growth rate of the economy as a whole.

The trend in consumption of oil and gas to 1980 will depend on the improvements that can be made in energy conversion efficiency, the rate of Soviet economic growth, and the influence on import-export policy of the balance of

Table 6.7 Western Analyses of Soviet Oil Export Potential in 1980 (million tonnes)

	1	*2*	*3*	*4*	*5*	*6*
1980 Production	607	640 (610-620)	640	640+	590	638
Imports	ns	ns	17.5	ns	15	ns
Availability	ns	ns	657.5	640+	605	ns
Demand	612/613	500	514	517	470	550
Export Availability	−5/−6	110+	143.5	123+	135	88

ns = not stated

Sources: 1. W Gumpel, *Sowjetunion: Erdöl und Nahostpolitik*, Aussenpolitik, 11/1971, p 677. Though Gumpel did not state a likely import level for 1980, he noted that the actual import-export pattern would be influenced by the need to maintain exports to Comecon for 'political' reasons. He did, however, state clearly (p 679) that by 1980 oil would have ceased to play a part in Soviet export trade. 2. Economist Intelligence Unit, *Soviet Oil to 1980*, London, EIU, 1973, p 33. The conclusion of the EIU analysis is somewhat unclear. The author stated that official Soviet sources (unspecified) estimated crude oil production in the Soviet Union in 1980 to be 610-620 million tonnes, but advanced the view that the likely level would be 640 million.
3. J Bethkenhagen, *Bedeutung und Möglichkeiten des Ost-West-Handels mit Energierohstoffen*, (Deutsches Institut für Wirtschaftsforschung, Sonderheft 104), Berlin, Duncker & Humblot, 1975, pp 61, 131, 253. 4. J Russell, *Energy as a Factor in Soviet Foreign Policy*, Farnborough, Saxon House, 1976, p 56. Russell did not give an estimate of imports: from his preceding analysis it appeared that the figure of 123 million tonnes constituted *net* exports. 5. E E Jack, J R Lee, H H Lent, 'Outlook for Soviet Energy' in *The Soviet Economy in a New Perspective*, Washington DC, US Congress, Joint Economic Committee, 1976, p 473. 6. A F G Scanlan, 'The Energy Balance of the Comecon Countries' in *Round Table on Exploitation of Siberia's Natural Resources*, Brussels, NATO, 1974, pp 97, 105.

Table 6.8 Western Analyses of Soviet Gas Export Potential in 1980 (billion cubic metres)

	1	*2*	*3*	*4*	*5*
1980 Production	460	400	400-440	390	454.5
Imports	13	15	15	15	ns
Availability	473	415	415-454	405	ns
Demand	430.7 (400)	365	363.3-402.3	346	412.5
Export Availability	42.3	50	51.7	59	42.0

ns = not stated

Sources: 1. Economist Intelligence Unit, *Soviet Natural Gas to 1985*, London, EIU, 1975, pp 45-47. The author of the report admitted (p 47) that his initial estimate of 1980 domestic demand may be only of the order of 400 billion cubic metres, permitting export availability to rise to 73 billion. 2. J Bethkenhagen, *op cit*, p 253. 3. J Russell, *op cit*, pp 33, 68, 70. 4. Jack, Lee, Lent, *op cit*, p 473. 5. Scanlan, *op cit*, p 97.

Table 6.9 Hungarian Estimate (1973) of Comecon Production and Demand to 1980

A. Oil Production and Consumption (million tonnes)

	Production			*Consumption*			**Production minus Consumption**
Year	*E Europe*	*USSR*	*Comecon*	*E Europe*	*USSR*	*Comecon*	*Comecon*
1969	16	316	332	51	262	313	19
1980	22	625-645	647-667	170	650	820	-153 to -173

B. Gas Production and Consumption (billion cubic metres)

	Production			*Consumption*			**Production minus Consumption**
Year	*E Europe*	*USSR*	*Comecon*	*E Europe*	*USSR*	*Comecon*	*Comecon*
1966	21.17	142.96	164.13	21.87	142.13	164.00	0.13
1968	26.75	169.10	198.65	28.30	168.87	197.17	1.32
1980	50.00	550-600	600-650	100.00	630.00	730.00	-80 to -130

Source: I Dobozi, *An energiahordozók a KGST gazdaságában*, Válóság, 1/1973, p 22

Table 6.10 Trends in Production and Domestic Availability of Oil and Gas 1971 to 1976

1. Oil (million tonnes, including gas condensate)

	1971	*1972*	*1973*	*1974*	*1975*	*1976*
Production	377	400	429	459	491	520
Imports (Oil and Products)	7	9	15	5	8	7
Availability	384	409	444	464	499	527
Exports (Oil and Products)	105	107	118	116	130	149
Domestic Availability	279	302	326	348	369	378
Growth Pattern	*1972 / 1971*	*1973 / 1972*	*1974 / 1973*	*1975 / 1974*	*1976 / 1975*	*1980P / 1980A*
Production (million tonnes)	23	29	30	32	29	139
(%)	6.1	7.2	7.0	7.0	5.9	28.3
Domestic Availability (million tonnes)	23	24	22	21	9	na
(%)	8.3	8.3	6.8	6.0	2.4	na

2. Gas (billion cubic metres)

	1971	*1972*	*1973*	*1974*	*1975*	*1976*
Production	212	221	236	261	289	321
Imports	8	11	11	12	12	12
Availability	220	232	247	273	301	333
Exports	5	5	7	14	19	26
Domestic Availability	215	227	240	259	282	307
Growth Pattern	*1972 / 1971*	*1973 / 1972*	*1974 / 1973*	*1975 / 1974*	*1976 / 1975*	*1980P / 1975A*
Production (billion cubic metres)	9	25	25	28	32	146
(%)	4.2	11.3	10.6	10.8	11.1	50.5
Domestic Availability (billion cubic metres)	12	13	19	23	25	na
(%)	5.6	5.8	7.9	7.6	8.9	na

A = Actual; P = Plan; na = not available

Note: the volume of Soviet exports and imports of oil and gas in 1977 was omitted from the 1977 foreign trade statistics. See A Nove, *The Times*, 18 August 1978, p 17.

Source: *Vneshnyaya torgovlya SSSR*, corresponding years

payments. The basic directives for the tenth Plan emphasize the need to use cheap solid fuel rather than oil and gas in electricity production, and set a target for reducing the amount of energy consumed in producing 1 kilowatt-hour to 325-328 grammes of standard fuel, compared with 340 grammes in 1975.[81] This objective was stressed again by Kosygin in his speech to the 30th Comecon congress in 1976, adding that industry ought to convert rapidly to less energy-intensive processes and that the construction of nuclear capacity should be accelerated.[82] The economy as a whole has exhibited decreasing energy-intensity and this is planned to continue. The target is a saving in the level of energy consumed of 10 per cent (150 million tonnes of standard fuel) of the energy consumption level derived by extrapolation to 1980 of the energy consumption to NMP coefficient of 1975.

It is evident from table 6.10 that in 1976 the rate of increase in the domestic availability of oil and products was only 2.4 per cent on the previous year, whereas availability of gas increased by 8.9 per cent. Taking into account the trends in consumption growth during the ninth Plan the annual expansion in the oil sector in 1976 compared with 1975 seems low, and could evidence some de-stocking to boost exports and domestic supply. Despite this, the trend is towards a declining annual rate of growth in oil consumption. Assuming that consumption increases by an average of 5.5 per cent per year between 1976 and 1980 and that imports reach no more than 10 million tonnes, the export surplus would be 135 million tonnes. This would be the case even if Soviet production in 1980 is at the lower limit of the initial target for the tenth Plan, which is 620 million tonnes. If oil production reaches the upper limit of 640 million tonnes the export surplus would be 155 million. Alternatively if one supposes from the 1971-1975 trend that a further decline of 0.5 per cent per year in oil consumption could be expected, and again assuming that imports total 10 million tonnes, the export surplus would be 170 million tonnes at 620 million tonnes' production, and 190 million at 640 million tonnes.

In view of the trends in production and consumption between 1971 and 1976 it appears that the extremes of the above range of figures are unlikely. The range of 155 to 170 million tonnes of oil and oil products available for export in 1980 is therefore suggested as a reasonable estimate, and the substantial growth in export volume in 1976 could be taken as an indication that the upper limit of the suggested range is attainable. Table 6.11 contains a detailed analysis of Soviet exports of oil and oil products in 1976, showing changes in volume and unit price compared with 1975. The unit price is, however, not strictly comparable on a country-to-country basis because of the effect of product mix and of the change in this mix over a period of time. Nonetheless it is unlikely that the mix has changed greatly in the major export markets where, given that refinery capacity is generally adequate to meet demand, the majority of purchases from the Soviet Union will be of crude oil. In all but a few cases there has been a continued rise in the unit price of Soviet deliveries. Unit prices in trade with Eastern Europe, however, have risen less than those charged to most of the other countries. In a few cases the unit price has dropped, but the volumes involved are small and a slight change in product mix could account for this, since the price levels suggest that refined products constitute the sum of the quantities involved.

The structure of Soviet trade in gas in 1976 is outlined in table 6.12. Total exports grew by almost 6.5 billion cubic metres, showing increases in deliveries to each of the established markets and the commencement of deliveries to France. As in the case of oil prices, the Soviet Union has taken advantage of the rising world price for gas in its trade with the West. This is also reflected in rising unit prices in trade with other Eastern European countries. In contrast to the position in the oil industry, domestic availability of gas has shown increasing rates of growth (table 6.10.2). However, the export level of Soviet gas to 1980 will depend on the success in meeting Plan targets and the rate at which it is used

Table 6.11 Soviet Oil Exports 1976 (Oil and Products)

Destination	*Volume (thousand tonnes)*	*Change cf. 1975*	*1976 Price (R/tonne)*	*Change cf. 1975*
Total Soviet Exports	148,514	+18,163	51.7	+6.4
of which: Oil	110,790	+17,720	46.2	+5.9
Products	37,724	+443	67.7	+9.8
(i) Comecon				
Bulgaria	11,868	+315	37.5	+3.7
GDR	16,766	+1,814	32.1	+3.9
Poland	14,073	+802	42.0	+2.5
Czechoslovakia	17,233	+1,268	34.1	+3.2
Hungary	8,435	+900	44.7	+3.7
(ii) EEC 9				
France	5,729	+2,422	65.0	+6.8
West Germany	7,132	-502	80.9	+18.7
Netherlands	2,674	-416	82.5	+17.5
Belgium/Luxembourg	2,082	+827	66.8	+3.7
Italy	11,982	+5,099	65.4	+8.4
Great Britain	4,051	+2,796	68.9	+4.2
Irish Republic	155	-21	54.5	+5.5
Denmark	1,632	+454	66.6	+9.0
(iii) Rest of World				
Austria	1,513	+186	65.5	+6.8
Greece	1,948	+60	67.0	+7.5
West Berlin	1,072	+224	74.4	+3.7
Iceland	417	-21	77.3	+3.8
Spain	2,002	+278	63.3	+5.7
Norway	218	-65	68.7	+5.2
Portugal	1,039	-17	64.5	+5.5
Finland	9,620	+912	66.4	+4.5
Switzerland	942	-18	71.6	+7.4
Sweden	2,729	-721	61.2	+11.1
Yugoslavia	4,858	+414	65.6	+4.3
Afghanistan	149	0	96.0	+9.4
Bangladesh	95	-72	85.3	+7.8
North Vietnam	439	+36	32.9	-2.4
India	1,113	-94	88.1	+10.6
Cyprus	257	+51	51.3	+6.4
North Korea	1,061	-49	41.2	+17.2
Mongolia	415	+51	61.8	-25.7
Nepal	77	+18	81.4	+8.9
Syria	385	+383	74.9	-308.1
Japan	1,773	+453	63.8	+12.6
Ghana	250	+106	64.7	-0.9
Guinea	81	+19	90.7	+15.4
Egypt	226	-5	89.8	+7.6
Liberia	24	-2.6	54.3	+2.8
Morocco	665	+16	64.2	+5.3
Somalia	136	+18	64.9	-7.7
Brazil	1,071	-404	65.6	-23.1
Canada	93	-127	64.4	+4.2
Cuba	8,809	+749	32.7	+1.9
United States of America	1,059	+520	65.5	+3.3

Source: *Vneshnyaya torgovlya SSSR 1976*, passim

Table 6.12 Soviet Gas Exports 1976

Destination	*Volume (billion cubic metres)*	*Change cf. 1975*	*1976 Price (R/thousand cubic metres)*	*Change cf. 1975*
Total Soviet Exports	25,780	+6,447	28.45	+5.11
(i) Comecon				
Bulgaria	2,229	+1,044	33.37	+3.98
Hungary	1.001	+0.4	33.95	+4.13
GDR	3,369	+67	27.71	+12.60
Poland	2,549	+40	32.00	+4.00
Czechoslovakia	4,287	+593	34.52	+9.05
(ii) EEC 9				
Italy	3,720	+1,378	14.00	-2.35
West Germany	3,976	+878	22.73	+4.90
France	993	+993	25.60	n app
(iii) Rest of World				
Austria	2,785	+902	33.45	+3.12
Finland	870	+151	47.26	+0.41

n app = not applicable

Source: *Vneshnyaya torgovlya SSSR 1976*, passim

as a domestic substitute for oil. There is no possibility of increasing imports from Iran and Afghanistan as the pipelines are currently operating near design capacity. After 1980, however, the pipeline system from Iran will be expanded. This has been decided under a recent agreement providing for the delivery of further quantities of Iranian gas into the Central Asian area of the Soviet Union, against the delivery of an agreed volume of Siberian gas via the 'Bratstvo' pipeline into West Germany. The Soviet Union will accept a part share of the Iranian gas as a transit fee.

Considering the past performance of the gas industry against the Plan it is difficult to estimate the likelihood of the target of 435 billion cubic metres being attained in 1980. It may, however, be indicative of Soviet confidence that the gas target for 1980 was eventually set at the upper limit of the original range. It may also be significant that in the early part of 1977, annual production targets to 1980 for the coal and gas industries were released, and that none was available for oil.[83] One possible explanation for this is that plans for oil production may be revised depending on the level of production achieved in coal and gas. If this proves to be high it would facilitate substitution of fuel oil by coal and of petrochemical feedstock and domestic fuels by gas. This would allow planners the choice of scaling down the oil production target for reasons of technical difficulty, high operating cost or for reserve conservation. Judging from the performance of the gas industry to 1977, there is a strong possibility that the target of 435 billion cubic metres can be attained in 1980. Even if a shortfall similar to that experienced in the ninth Plan was repeated, the production level is hardly likely to be less than 400 billion cubic metres. If, therefore, imports remain at 12 billion cubic metres and, assuming that domestic consumption was to grow by 8.5 per cent per year between 1976 and 1980 (this is only slightly below the rate of increase in 1976 over 1975), then an export surplus of 31 billion cubic metres would be available in 1980, if production then was 435 billion. If, however, consumption was to grow at an average of 6 per cent per year after 1976, an export surplus of 59 billion cubic metres is virtually certain. In the light of the gas industry's performance to mid-1977 the latter figure is considered realistic.

The need to import grain and technology from the West during the ninth Plan accounted for the substantial growth in Soviet imports. Export prospects deteriorated as the developed Western world became seriously affected by inflation and economic recession. Despite this, the enhanced value of fuels exports accounted for much of the increase in the Soviet Union's non-Comecon trade, and these larger revenues were most needed after the grain purchases of 1972 and 1973. As Philip Hanson stated,[84] the Soviet Union's trade with non-Comecon countries during the ninth Plan was some 50 per cent above the level implied in the original directives. Although it is likely that there will be some deceleration in imports, other than those of grain, during the tenth Plan, there is still a large debt to service. Hanson and many others believe this debt will grow up to 1980. Accordingly the increased export of fuels and energy, especially of oil and gas, may be unavoidable, to the extent that availability for domestic and Comecon deliveries may be determined only after taking into account the demands of the deficit on trade with the West. As shown in the 1976 trade figures, the Soviet Union is expanding its oil trade with the West and is developing its gas trade with Western Europe particularly rapidly. The income from oil and gas trade will not, however, clear the Soviet debt: it may serve to keep the Soviet Union from the point at which international institutions might call a halt to the extension of credits.

References

1. B Askanas, H Askanas, F Levcik, 'Structural Developments in CMEA Foreign Trade over the Last Fifteen Years (1960-1975)', *Forschungsbericht*, 23/1975, Wiener Institut für Internationale Wirtschaftsvergleiche, pp 3-7.

2. S Albinowski, in *Polityka*, 24 September 1966. An English translation of this article appears in R E Ebel, *Communist Trade in Oil and Gas*, New York, Praeger, 1970, pp 249-254. Subsequent references relate to this translation.

3. *ibid*, pp 250-251.

4. W Gumpel, 'Sowjetunion: Erdöl und Nahostpolitik', *Aussenpolitik*, 11/1971, pp 670-681.

5. *ibid*, p 676.

6. N V Melnikov, 'Voprosy razvitiya toplivnoi promyshlennosti', *Planovoe khozyaistvo*, 1/1969, p 12.

7. Gumpel, *op cit*, p 677.

8. *ibid*.

9. *ibid*, pp 679-680.

10. *ibid*, p 680.

11. *ibid*, p 681.

12. Albinowski, in Ebel, *op cit*, p 252.

13. Gumpel, *op cit*, p 681.

14. Ebel, *op cit*, p 107.

15. *ibid*, p 109.

16. *ibid*, p 111 (based on forward estimates given by the Soviet Oil Minister, V D Shashin, in 'Neftedobyvayushchaya promyshlennost' SSSR za 50 let Sovetskoi vlasti', *Neftyanaya promyshlennost'*, 10/1967, p 7).

17. *ibid*, p 119.

18. Editorial, 'Na blago rodiny', *Stroitel'stvo truboprovodov*, 5/1968, p 7.

19. *Stroitel'naya gazeta*, 5 February 1969, p 1.

20. *Bakinskii rabochii*, 13 March 1969, p 2.

21. Ebel, *op cit*, p 129.

22. V G Vasil'ev, 'Gazovaya promyshlennost' – detishche Oktyabrya', *Geologiya nefti i gaza 1967*, No 12, p 1.

23. Ebel, *op cit*, p 134.

24. *ibid*, p 135.

25. *ibid*, p 166.

26. Askanas, Askanas, Levcik, *op cit*, p 26.

27. *ibid*, p 29.

28. *ibid*, pp 38-39.

29. M C Kaser, 'Technology and Oil in Comecon's External Relations' (Review Article), *Journal of Common Market Studies*, 13/1975, pp 161-172.

30. *ibid*, pp 166-167.

31. *ibid*, p 171.

32. *ibid*, p 172.

33. M C Kaser, 'Oil and the Broader Participation of IBEC', *International Currency Review*, 6/1974, pp 25-27, 32.

34. *ibid*, p 25.

35. J Bethkenhagen, *Bedeutung ung Möglichkeiten des Ost-West-Handels mit Energierohstoffen*, (Deutches Institut für Wirtschaftsforschung, Sonderheft 104), Berlin: Duncker & Humblot, 1975.

36. *ibid*, p 266.

37. *ibid*, pp 189, 294.

38. *ibid*, p 298.

39. *ibid*, p 299.

40. *ibid*, pp 190, 295.

41. *ibid*, p 204.

42. *ibid*, p 253.

43. *ibid.*

44. A W Wright, 'The Soviet Union in World Energy Markets', in E W Erickson, L Waverman (eds), *The Energy Question: An International Failure of Policy*, Toronto, University of Toronto Press, 1/1974, pp 85-89. (Note also the existence of an earlier work by this author, his unpublished PhD thesis entitled *The Theory and Practice of Soviet Investment Planning with Special Reference to the Mineral Fuels Industries*, Massachusetts Institute of Technology, 1969.)

45. *ibid*, p 85.

46. *ibid*, p 91.

47. *ibid*, p 94.

48. *ibid*, p 95.

49. P R Odell, *Oil and World Power: Background to the Oil Crisis*, Harmondsworth, Penguin, 1975 (4th edition), pp 47, 49.

50. M I Goldman, *Detente and Dollars: Doing Business with the Soviets*, New York, Basic Books, 1975, p 17.

51. J Wilczynski, *The Political Economy of East-West Trade*, New York, Praeger, 1969, p 170.

52. R Portes, 'West-East Capital Flows', occasional paper, University of London, Birkbeck College, March 1977, cited in *Financial Times*, 5 May 1977, p 32. Portes estimates an increase in the Soviet debt to the West from 5.9 billion dollars at the end of 1974 to 11.4 at end 1975 and 14.4 at end 1976.

53. A Nove, *The Soviet Economy*, London, Allen and Unwin, 1969, pp 214-215.

54. A Zwass, *Monetary Cooperation between East and West*, New York, IASP, 1975, p 46.

55. Nove, *op cit*, p 217.

56. Zwass, *op cit*, p 50.

57. *ibid*, pp 59-60.

58. *ibid*, pp 144-146. The principle is to arrive at a world market price by eliminating seasonal, speculative and monopolistic elements over an extended period and to fix these for the duration of a given Five-Year Plan. For example, basic prices for the 1971-1975 Plan were calculated on the average of those for 1965-1969. World market prices were then expressed in an appropriate Western convertible currency and converted to the Comecon equivalent at the current dollar-to-ruble exchange rate. See also O I Tarnovskii, N M Mitrofanova, *Stoimost' i tsena na mirovom sotsialisticheskon rynke*, Moscow, Nauka, 1968, p 84.

59. Zwass, *op cit*, p 146.

60. *ibid*, p 65.

61. Zwass, *op cit*, p 152.

62. *Trybuna ludu*, 8-9 February 1975, p 5.

63. *Rudé právo*, 30 March 1974, pp 1-2; *Financial Times*, 2 April 1974, p 4.

64. *Revista economică*, 12/1975, pp 9-10.

65. *Financial Times*, 16 January 1975, p 6; 24 February 1975, p 8.

66. *Financial Times*, 20 March 1975, p 6.

67. *Financial Times*, 11 July 1974, p 8; 19 August 1974, p 5.

68. *Petroleum Economist*, March 1974, p 99.

69. Zwass, *op cit*, p 57.

70. A Nove, in *The Times*, 17 July 1974, p 21.

71. See B Dasgupta, 'Soviet Oil and the Third World', *World Development*, May 1975, p 358.

72. *Financial Times*, 26 February 1976, p 4.

73. *The Times*, 10 September 1975, p 4; 20 September 1975, p 4.

74. *International Herald Tribune*, 2 October 1975, p 8.

75. J Russell, *Energy as a Factor in Soviet Foreign Policy*, Farnborough, Saxon House, 1976, p 178.

76. *Financial Times*, 28 January 1975, p 14.

77. Bethkenhagen, *op cit*, p 208.

78. I Dobozi, 'An energiahordozók a KGST gazdaságában', *Válóság*, 1/1973, pp 18-27.

79. See A F G Scanlan, 'The Energy Balance of the Comecon Countries', in *Round Table on the Exploitation of Siberia's Natural Resources*, Brussels, NATO, 1974, pp 83-107. He argued (p 96) that demand for energy in the Soviet Union would rise at an average of six per cent per year between 1972 and 1980, and for oil at a rate of eight per cent per year, reaching 550 million tonnes of standard fuel in 1980. This rate of growth in total energy demand is the same as that of the previous decade.

80. R W Campbell states, 'without trying to project specific rates, we might still conclude that there would be some acceleration of the growth of consumption in the coming period compared to the sixties, and as the rate of growth of primary energy production is unlikely to accelerate, the prospect is for heavy pressure to reduce exports'. 'Siberian Energy Resources and the World Energy Market', in *Round Table . . ., op cit*, p 79.

81. *Izvestiya*, 14 December 1975, p 2.

82. A N Kosygin, 'Ocherednye zadachi sotsialisticheskoi integratsii', *Planovoe khozyaistvo*, 9/1976, p 5.

83. Annual targets for coal production to 1980 are given and discussed in *Ekonomicheskaya gazeta*, 4/1977, pp 1-2, and for gas production in *Ekonomicheskaya gazeta*, 6/1977, pp 1-2. However, the discussion of oil developments given in *Ekonomicheskaya gazeta*, 8/1977, p 2, does not detail annual production targets.

84. P Hanson, 'USSR: Foreign Trade Implications of the 1976-1980 Plan', *Economist Intelligence Unit Special Report*, 36/1976, London, EIU, 1976, p 3.

Chapter 7

Soviet Relations with Oil and Gas Producers

As has been shown in earlier chapters the Soviet Union is self-sufficient in energy, including oil, and has the capacity to supply the major part of Eastern Europe's energy needs. It is, at the same time, able to maintain a level of exports to the West adequate to support essential imports and to help avoid the risk of a balance of payments deficit, such as would prompt the industrialized West into limiting the extension of low-interest loans facilitating such imports. In view of the foregoing it is reasonable to question the basis for any interest by the Soviet Union in negotiating for supplies of oil and gas from other producers. However, imports can serve other than domestic needs, and hydrocarbon trade in particular reflects to a considerable degree the coalescence of economic, political and strategic elements of Soviet relations with the rest of the world, and particularly with the US. Discussion of the Soviet Union's apparently growing need for Middle East oil from the early seventies, pointed to an impending change in superpower relations since it was thought that the Soviet Union could not long maintain self-sufficiency in energy. However, the events of 1973-1974 necessitated considerable rethinking of the role of energy, especially of oil, not only in respect of the import needs of much of the industrialized world, but also of the interrelationship of the major powers.

A nation that is dependent on unstable foreign sources for its energy supply is in danger of having that supply interrupted at any time. A country that has relied on a supply of OPEC oil in the pre- and post-Yom Kippur period has been obliged to restructure its foreign trade, and generally needs to spend considerably more on the import of oil. Also its terms of trade with OPEC countries will have deteriorated. Moreover, given the fear of possible further interruption of supply, a highly dependent net importer has experienced the narrowing of foreign policy options, especially in the Middle East. In contrast, the nation enjoying energy autarchy, if necessary at a tolerable cost, can remain aloof from the vacillations of an unstable international market. Such a nation also enjoys a wider range of options in external relations, enhanced by the possession of a vital resource such as oil. As long as a surplus over domestic demand is maintained this would afford a measure of logistic freedom, from which would stem greater flexibility in trade strategy and foreign policy.

The multidimensional nature of the Soviet role in the Middle East, the world's major oil-producing area, has been highlighted in recent publications by Ian Smart and Hannes Adomeit. Ian Smart stressed four principal interacting influences that shape the policy of the US and the Soviet Union with regard to the

Middle East: first, that the area itself is an arena of violent international conflict on frequent, though generally unpredictable occasions; second, that the two countries conventionally termed 'superpowers', namely the US and the Soviet Union, have varied interests and influences in the area; third, that each superpower is concerned with the international energy market, though each to a different extent and in a different way; and fourth, that the conduct of policy in the Middle East on the part of each superpower affects their own economic and political interactions.[1] Concentrating exclusively on the question of the Soviet role in the Middle East, Hannes Adomeit pointed to three major elements shaping Soviet policy: first, that the Soviet Union regards as illegitimate the setting-up and continued existence of the state of Israel in Palestine, the major cause of conflict in the area; second, that the Soviet Union attaches great importance to its ability to intervene militarily on behalf of states perceived to be friendly; and third, that the Soviet Union believes it has an ideological commitment to assist the Arab nations in freeing themselves from a colonial past.[2]

To understand the nature of the Soviet Union's interest and involvement in Middle East oil it is necessary to look not only at the course of Soviet policy in the area prior to 1973, but also at the process of development of Arab moves to independence in oil affairs, which were catalyzed through OPEC. OPEC rightly judged the moment at which oil could be employed as an element of general strategy and confirmed a fundamental change in international economic and political relations. This last point is particularly important in any analysis of Soviet, American and Arab interrelations in the post-Yom Kippur War period.

Adomeit indicated a major analytical problem, namely that while it can be said that governments act in accordance with what they believe to be the 'national interest', this concept is neither predetermined nor immutable.[3] The concept of acting in the 'national interest' may be applied as the retrospective validation of a corporate reaction to events that is little more than the result of the uncoordinated interaction of conflicting influences.[4] The Soviet Union has made stringent efforts to increase military, political and economic influence in the Middle East. It is, however, extremely difficult to measure quantitatively or qualitatively the effects of such influence as might be reflected in, for example, changes in economic or governmental systems of those countries in which a Soviet presence is detected.

Furthermore it is misleading to assume the existence of an overall coordinated strategy on the part of the Soviet Union, drawn up and directed by the Party's Politburo. The detectable gulf between Soviet pronouncements and actions also serves to complicate the analytical process.[5] In a study completed very soon after the Yom Kippur War the American analysts Kohler, Gouré and Harvey drew the picture of an aggressive Soviet Union showing 'little evidence of willingness to cooperate in the preservation of international stability in general or for the attainment of a settlement in the Middle East in particular, except on Soviet terms'.[6] They did admit that the degree to which the Soviet Union could influence Arab states to pursue policies optimal to Soviet interest was limited, but that opportunities existed for disruption by the Soviet Union. This would make more difficult any negotiations aimed at securing a settlement. Why the Soviet Union would wish to do this is not made clear.

In the case of the US it has been pointed out that popular opinion was not in favour of American involvement in any Arab-Israeli conflict. Also in the aftermath of the 1967 six-day war the proportion of the American nation sympathetic to the Israeli cause was declining, but Congress was still committed, by a substantial majority, to support Israel, believing that the 'national interest' lay in countering (perceived) Soviet designs in the Middle East. This despite the opposition of a large majority of the population to the continued supply of arms to Israel.[7] Up to the commencement of the Yom Kippur War few members of Congress appreciated the potential danger of such a policy. One American analyst, J C Campbell, has pointed out that a fundamental error of judgement was made in underestimating the degree of cohesion that would be exhibited by the Arabs, acting through OPEC. Consequently the forces that were to determine the course of events and the scope for negotiations in the aftermath of the war could not be accurately gauged.[8] This policy of support for Israel, whilst 'containing Communism', led to a measure of economic loss through destruction of the balance in negotiations between the oil companies and OPEC over the future supply and price of oil. These factors, plus the cost of a failed policy in South-East Asia have led, in the view of the British analyst Douglas Evans, to acceptance by the US of a diminished role in world politics.[9]

This role is, however, open to influence by the home-based oil industry, whose sphere of activity transcends national boundaries. As a result of the growth of the multinational oil industry, the potential for a conflict of interest arises between companies and governments of the home and host country. Dependent on mutual perception of the alternatives open to each, a range of policy options can be determined. Thus acting within the framework of national and international law an oil company can formulate and put into effect policies which are financially attractive to themselves, but which may run counter to the home government's preference. J S Nye has put forward the view that any influence possessed by the oil companies is characterized by their playing either a direct role in executing a privately-determined foreign policy, or an indirect role in influencing the course of intergovernmental relationships.[10] There is ample evidence of the effect of either role as played by multinational companies. However, as far as the course of political developments in the Middle East is concerned, Walter Laqueur has advanced the view that 'in spite of their great economic significance the oil companies have not during the last decade acted from a position of political strength', and that they have been able 'neither to provoke wars nor prevent wars', since 'there has been a striking discrepancy between their economic significance and their political power'.[11]

Although the frontiers of a state define the extent of its right to command, it is unrealistic to expect complete control within these frontiers. The more a state tends towards pluralism, the more it has to strive to ensure acceptance of its corporate will. What is distinctive about the relationship of the oil companies and the American government is that persistent government attempts at resource allocation have repeatedly to be revised because the implementation of policy runs counter to the interest of those called on to implement it. The political success recorded by gas consumers in the US, in keeping domestic prices at an artificially low level, deterred investment in the industry at a time when the environmentalist lobby prompted restrictions in open-cast mining and coal burning in densely populated areas. Hence the US became substantially dependent on imported energy in the form of OPEC oil and this dependence shows no sign of diminishing despite the price rises and 'Project Independence 1980'.

The foregoing is complicated by the changing role of the oil-producing states themselves, not only in international politics but also in the world energy market. Account must be taken of this factor in discussing oil in relation to Soviet policy in the Middle East and to Soviet dealings with other powers, notably the US. This is not to suggest that through OPEC the producer states may be regarded as a homogenous group, with agreement amongst members on a unified policy of economic and social development. Indeed, the very diversity of objectives within the producer states has been internally problematic. There are wide social, economic and political differences which determine the preferred policy of the individual states and which serve to illustrate why their actions have often not been unified.[12] It is also worth noting that as far as the hostilities of 1973 were concerned the countries immediately involved, namely Syria and Egypt, were not members of OPEC, but were ethnically identifiable with the major oil producers. However, as the Minister for Petroleum of the United Arab Emirates has pointed out, the hostilities gave the oil-producing Arab states the opportunity of enhancing the influence of Arab interest in general.[13] The war itself helped to make legitimate a pricing and output policy that had been debated within OAPEC, of which Egypt and Syria were members, but not thus far in OPEC.[14]

One can detect two major causes of tension which influence OPEC's developing view of the world petroleum market. The latter half of the sixties and the early seventies were characterized by inflation in most of the industrialized world, on which the Arab countries depended for capital goods and technology to raise them from a comparatively low standard of living. In the US, for example, the Consumer Price Index had risen by 23 per cent between 1965 and 1970 and rose a further 20 per cent between 1970 and mid-1972. Similar or even larger increases were recorded in many European countries, and in Japan. Whereas this affected goods and services essential to the Arab countries, the price of oil, as indicated in Chapter 1, fell in real terms up to 1970, and it was only after the Libyan success in unilaterally raising oil prices that the prospects for action, independent of the oil companies, appeared possible. The second influence was the growing desire of the Arabs for national ownership of oil production and distribution facilities. This was not envisaged as an element of the dialogue between the companies and host countries which were pursued under the auspices of the London Oil Policy Group. It was initially an issue separate from discussion on prices, as far as Arab producers were concerned. However, the two issues coalesced in the events of 1973-1974, in the aftermath of which, previous agreements relating to the gradual relinquishing of concessions by the companies to the host governments were abandoned. At this time moves were started for a complete takeover by the host governments of the management of the producing companies.[15]

Though signs of disagreement within OPEC have been manifest on the question of pricing, notably in the negotiations leading to the price rise of December

1976 when members imposed different percentage increases,[16] the growth in the financial strength of OPEC is not in doubt. It is appreciated in the Organization that there are continued benefits to be derived from solidarity and that there are dangers in bringing about too rapid a substitution of oil by other fuels. The principal feature of the oil market after 1973 is that OPEC does not *need* to look for additional outlets, whether in Japan, Western Europe, the US or Comecon. Its common objective is that of retaining the means of controlling the development of the world petroleum market, and in so doing it is conditioned to the impact of such external factors as the possible American escalation of military and economic aid to Israel. In the case of the Soviet Union, as a potential market for Arab oil, the immediate issue is that of using the one product the Arabs have in free supply to offset the debt arising out of the Soviet Union's supply of arms. However, the need to deplete reserves in order to do this is falling as the Arabs are able increasingly to pay in hard currencies. It should also be borne in mind that the biggest arms deals between the Soviet Union and Middle East countries have been negotiated with those countries who cannot offer oil in return, namely Egypt and Syria.

A considerable volume of published work exists on the interrelationships of the Soviet Union, the US and the Middle East. A critical examination of this is beyond the scope of this volume: however, the issues raised in a number of these works are worth quoting in order to outline the mainstream of Western analysis on what is now regarded as a major problem of interstate relations in the seventies.

Writing in 1969, Walter Laqueur expressed the view that the Soviet Union's interest in Middle East oil was confined to propagandist attacks on the activities of (primarily) American companies operating in the area. He also maintained that there was little basis for the hypothesis that the Soviet Union was seeking 'control' over oil as a potential means of denying supply to Western Europe.[17] He stressed the significance of the year 1966 as a turning point in Soviet policy, following the agreement with Iran for the supply of natural gas. This was the first sign of Soviet need for a non-indigenous fuel. Whilst indicating that political considerations may in general be an element in the formulation of Soviet policy, Laqueur pointed out that as the cost of production in the Soviet Union is often greater than the import cost of a product, the purchase of Middle East oil or gas in no way differs from the import of any other commodity. The decision is justifiable, he maintained, solely on economic grounds. He regarded as more significant the Soviet Union's need to accept oil as the balancing element of barter trade.[18] The circumstances surrounding the possible role of oil as an element of trade between Eastern Europe and the Middle East are somewhat different. The latter's debt, incurred in respect of arms trade, is not as high as with the Soviet Union, and the capacity of Middle Eastern countries to absorb goods produced in Eastern Europe, is relatively low. Writing in 1969, the American analyst Stanislaw Wasowski indicated that if Eastern Europe were to attempt to satisfy its additional requirement for oil by supply from the Middle East, goods worth between one and two billion dollars at 1966 prices would have to be exported.[19] Laqueur had previously maintained that this figure was well beyond the Middle East's absorption capacity.[20] Though he did not offer an analysis of supply and demand for oil and gas in the Soviet Union, Laqueur noted that Middle East hydrocarbon resources were gaining importance for the Soviet Union without becoming a policy-shaping factor in the area, since the country had thus far been able to afford a policy of energy autarchy which was likely to continue.[21]

In a Soviet work published at approximately the same time as Laqueur's, L Z Zevin confirmed that the import of Middle East oil had become, in certain instances, economically advantageous for the Soviet Union.[22] In a further work of that period N P Shmelev indicated that the agreements reached between the Comecon countries and Middle East oil producers represented only the first steps towards further trade.[23] Zevin likewise stated that cooperation in the development of natural resources, with a view to their possible export to Comecon, was only in its initial stages.[24] Zevin was aware of the constraints on the ability of Comecon member countries to enter into such agreements with Middle Eastern producers, namely the provision of exportable goods sufficient to compensate for increased imports, in view of the tight domestic supply.[25]

Bearing in mind Laqueur's analysis and the developing Soviet view, two analyses of oil as an element of Soviet policy in the Middle East, which were written in the early seventies, attempted to determine policy options based on a range of supply and demand balances.[26] W Gumpel's broad hypothesis was that the Soviet Union's desire to maintain self-sufficiency supersedes all other considerations. He stated further that policymakers would on balance be prepared to sustain economic loss in its pursuit, but that if this loss (unspecified) were to become excessive, then the best solution to an overall energy deficit would be the import of oil from the Middle East.[27]

In contrast Sabine Baufeldt argued that the Soviet Union's net surplus of oil would decline for economic reasons alone. There would be three possible effects arising from this projected decline: first, that exports to the West would decline, the rate depending on growth in domestic consumption, thus necessitating greater dependence on the Middle East; second, that the Soviet Union could maintain exports to the West and to fellow members of Comecon only by substantially increasing imports from OPEC; third, that Soviet exports to the West would be maintained but that Eastern Europe rather than the Soviet Union could increase its imports from OPEC.[28] On the grounds that the Soviet Union needs technology from the West and that oil and petroleum products are a particularly useful way of securing such goods, Baufeldt ruled out the first option. The second option, she argued, was unlikely to be adopted on the grounds that barter arrangements could not be stretched to cope with the payments required. She regarded the third option as the most likely, stating that if the total Comecon deficit could be kept to 40 million tonnes per year then agreements concluded up to 1973 would be adequate.[29]

Concerning Soviet desire to ensure security of supply in the (in her view likely) event of import dependency, she outlined three possible Soviet strategies in relations with Middle East producers. First, she suggested the possibility of direct collaboration with producing companies in the Middle East, that is to say that the Soviet Union might fulfil a role broadly similar to that of the major oil companies, enabling the Soviet Union to deliver its allocation of oil produced to established Western and Eastern European markets, or in the case of an embargo on supplies to the West to withhold it. It is, however, argued that this option is likely to be constrained by Arab unwillingness to countenance collaboration with the Soviet Union any more readily than with American (or any other) enterprises. On another level – the Soviet Union might be wary of undertaking financial commitments in Arab countries following the costly experience of Egypt, which came to a head in 1972 with the expulsion of Soviet advisers. Moreover, the problem would arise of reconciling direct Soviet involvement in production with the stand taken by the Soviet Union supporting the right of the Arab nations to have unfettered control of their own resources.[30] The second option was the Soviet use of political (unspecified) influence to prompt producers to cut back supplies to the West, although at the time of writing the author stressed the ability of the Arab states to take decisions independently of either of the superpowers, and that the Soviet Union might therefore be restricted to a post-factum support of Arab policy.[31] The third option, in Baufeldt's view the most likely, was described as 'indirect influence', the possible manifestation of which would be overt support for Arab moves towards nationalization. This might be effective as an element of strategy in the short term, since it would help to secure a rising income for oil producers and diminish US influence through the oil companies. Both these factors are in the interest of the producers and the Soviet Union.[32]

This last view contrasts somewhat with that expressed in an earlier study by Robert Hunter.[33] Accepting the view that the Soviet Union was about to become an oil importer, Hunter argued that the change in status to net importer might come as early as 1975.[34] He ruled out the possibility of Soviet 'colonialist' intervention in the Middle East, stating that the Soviet Union would rather seek supply through normal commercial arrangements.[35] The Soviet Union's objectives, he argued, would be better served by adopting a 'temperate attitude towards relations between the oil-producing states and Western oil companies', so as to avoid cuts in Middle East oil revenues which would be detrimental to the Soviet Union.[36] However, Hunter's view of the role of oil in Soviet policy in the Middle East was somewhat inconclusive. He saw the problem as an unresolved dilemma affected by the difficulty of making a decision when faced with the abandonment of what he believed to have been a fundamental Soviet objective since the time of the Revolution, namely self-sufficiency in oil.[37]

A further perspective was offered by Abraham S Becker in a paper written in 1973.[38] His was the earliest analysis to stress that the issue of whether the Soviet Union would wish to influence events in the Middle East by aiming for stability in the area, or by promoting radical nationalist militancy, overlooked the emergence of the Middle East, and especially the Arab states, through OPEC, as an independent force in world politics.[39] He shared the widely-held view of the Soviet Union's impending oil deficit (though he did allude to the possibility of substitution by gas[40]) largely on the grounds that the development of Siberia would take decades, rather than years.[41] In conclusion he stated that whereas the net export position of the (Comecon) group would be under no threat in the next few years, by 1980 an overall CMEA deficit was likely.[42] It is difficult to comprehend Becker's view of the likely time-scale of Siberian energy development when it is borne in mind that the basic directives for the ninth Plan were available when his analysis was being written, and that these clearly outlined the importance of the region for total production.[43] Also neglect of the changing role of natural gas detracted from his analysis of supply and demand for oil. Nevertheless he did highlight an important point in stressing that the economic motivations for the Soviet Union's possible increase in imports from the Middle East should not be overdrawn, since the classical motivations had been political.[44] Arguing that the Soviet Union was not averse to economic warfare in principle, he noted that a future oil crisis might see Moscow enthusiastically backing radical measures undertaken by the producer states depending on the breadth of the producer coalition and the objectives at stake.[45]

The crystallization of Becker's thoughts on the Soviet role in the Middle East is found in a later work of his. A further dimension introduced into the discussion was that of the interaction between Soviet policy in the Middle East and relations with the US.[46] With hindsight Becker stressed that the Soviet Union had been cautious about undertaking direct commitments to support any one country, or any group of Middle East states, and that pronouncements on policy had been confined to the safety of anti-Zionism and anti-oil imperialism.[47] Aware of the fierce internal rivalries for supremacy within the Arab world, the Soviet Union, he argued, had developed the skill necessary to support, for example, Iran and Iraq simultaneously: the rivalry between Baghdad and Tehran for supremacy in Arab affairs had not so far required the Soviet Union to take the side of either party.[48]

A somewhat different view of the relevance of oil to Soviet policy in the Middle East is provided by the American analyst Lincoln Landis.[49] He argued that the main features of Soviet energy planning were the opportunity provided for developing all forms of energy resources domestically, increasing demand in domestic and export markets, and a deficient planning system, unable to cope with short- and long-term requirements.[50] He saw Middle East oil as having a 'compensating and corrective' function with the Soviet Union becoming increasingly involved in Middle East oil affairs, developing the role of 'strategic middleman'.[51] His view was that the Soviet Union would be importing quantities of oil to compensate for domestic shortcomings whilst seeking to maintain developing export business in existing markets. However, Landis extended his theory of the Soviet Union as a 'strategic middleman' by postulating the ultimate objective of a Soviet-controlled 'world energy delivery system within a world socialist planned economy'. This would enable the Soviet Union 'according to its own choosing to continue in the use of politics and to exert pressures upon capitalist states by threatening their strategic interests, which include the unhampered flow of petroleum from the Middle East'.[52] Though the extention of Landis' theory might seem exaggerated in view of the considerable constraints that would be imposed on such a policy, not only by the Middle East producers themselves but by the Western states and institutions who would see their interests threatened, the concept of 'strategic middleman' is useful in looking at the Soviet role, and sheds some light on the degree of flexibility within the Soviet energy planning process.

It is, however, only in a very recent study that the key issue is highlighted and discussed, namely the limitations on policy in the Middle East imposed by considerations of detente between the Soviet Union and the US, with the policy vulnerable in the event of a military confrontation.[53] Galia Golan argued that the level of cooperation between the Soviet Union and the Middle East producers varied in their attitude to the Soviet Union from neutrality to outright hostility.[54] Hence, she concluded, the Soviet Union was still far from able to influence the use of the 'oil weapon' in its own interest. Equally important, as she pointed out, the rate of expansion of the Soviet share of Western oil markets scarcely constituted competition for the Arab producers.[55]

However, Arthur Jay Klinghoffer argued that the influence of the Soviet Union lies not so much in the extent of its need to import oil, either now or in the future, but in its ability to influence the policy of producers towards the governments of political adversaries via the companies which are commercially involved in the area. He admitted that in some instances the Soviet interest is economic, in that the import of Middle East oil could be cheaper than the development of marginal Siberian fields. In his view the prime motive was to limit the extent to which the US could finance the military build-up of Israel, via the profits accruing to those who, as the Soviets saw it, represented the interests of Zionist capital.[56]

The foregoing analyses of Western views of the Soviet Union's role in the Middle East is intended to emphasize the diverse factors that influence the formulation of policy. It is not assumed that in the end behaviour is determined in accordance with a fully coordinated and optimized decision-making process. Indeed, as Hannes Adomeit pointed out,[57] the scope of the Soviet Union's international relations had become so vast as to render impracticable any attempt at a unified 'decision theory' model, aspiring to weigh each variable by accepted objective criteria. However, given the existence of a Soviet presence in the Middle East and the conflicting perspectives of Soviet interest in its hydrocarbon resources found in Western discussion of the future development of Soviet energy, an attempt must be made to examine the involvement of the Soviet Union as it relates to the economic dealings with Middle East oil producers: and to extend this to consider Soviet relations with other producers, bearing in mind the changing dynamics of the world petroleum market.

The Soviet Union and the Middle East

In a trading relationship the imbalance between one country's need to export and another's to import, affords one party a measure of advantage which may be used to influence matters related or unrelated to the trade agreement. Even if the supplier or recipient with the advantage chooses not to exercise that influence, the possibility of this happening is likely to affect the perceptions held by each country of the 'rules' or scope for negotiation that govern their relationship. Specifically, whilst Soviet trade with Middle East countries over the period 1971-1975 constituted approximately 7 per cent of its total world trade, the case of trade with the Soviet Union, from the point of view of certain Middle East countries, is completely different. For example, in 1974 some 23 per cent of Syria's exports were destined for the Soviet Union.[58] In the case of Egypt, approximately 28 per cent of its exports were directed to the Soviet Union between 1960 and 1974, this figure rising to almost 40 per cent if other Eastern European countries are included.[59] In each case there was a considerable burden in that exports need to be maintained in order to repay the cost of development, goods, and expertise previously provided by the Soviet Union, this in itself constituting a constraint on flexibility in trade policy. In particular, the burden of Egypt's debt to the Soviet Union for arms supplies has been such that around 50 per cent of its principal exportable commodity, raw cotton, has effectively been mortgaged to the Soviet Union.[60]

Table 7.1 details the visible trade balance between the Soviet Union and the principal oil producers, predominantly Arab Middle East states, from 1971 to 1976; the importance of oil and gas in the total import trade of the Soviet Union is given in table 7.2, and absolute volumes and prices in table 7.3. It will be seen that there was a substantial drop in imports of oil in 1974, compared with 1973, following OPEC's quadrupling of prices. This factor in itself casts some doubt on the accuracy of earlier views of the unavoidable need of the Soviet Union to increase imports. The slight change in the relative importance of oil in the Soviet visible trade balance was sufficient to alter the position from surplus to deficit. Sustaining only a slight drop in total exports in 1974, compared with 1973, (see table 6.2a), the Soviet Union was able to maintain growth in net exports in 1974 and to record further gains in 1975 and 1976. It is possible of course that in 1974 some run-down of stocks took place as plans were reformulated to take into account the increased import price and the enhanced opportunities for earning hard currency in non-Comecon export markets afforded by OPEC's action. Another possibility, though less likely, is that the high figure of 13.2 million tonnes for imports of crude oil from the Middle East in 1973 reflected overbuying and stockpiling in anticipation of the price increase, this resulting in lower demand for imports in 1974 and the resumption of the upward trend in 1975 and 1976, at levels only slightly below those that might have been anticipated. Such a hypothesis assumed that the Soviet Union was able to anticipate correctly the extent and timing of the price rises. Though the Soviet Union undoubtedly did support the Arab use of the 'oil weapon' in the form of a supply embargo, and likewise Arab attempts to raise prices,[61] the problem in determining the 'Soviet view' is that of distinguishing between what the Soviet Union felt capable of influencing, and what was regarded as inevitable.

Soviet reaction in cutting direct imports of OPEC oil following the price rises of 1973 and 1974 and at the same time largely maintaining a trade balance with the two major partners, Iraq and Iran is shown in table 7.1. The position of gas imports is different. Whereas the unit import price for Iranian and Afghan gas remained stable between 1969 and 1973, prices rose sharply in 1974 and 1975.[62] The price rise for Afghan gas was negotiated in June 1974 and made effective from 1 October that year;[63] that for Iranian gas was negotiated in August 1974 and became effective retroactively from 1 January 1974.[64] The Soviet demand response to these changes was the opposite of that following the oil price rises. In view of the absence of an alternative market for Afghan and Iranian gas it might seem hard to account for the success of these negotiations.

A probable explanation is that counterpricing of goods imported from the

Table 7.1 Soviet Visible Trade Balance with Principal Oil and Gas Producers 1971-1976 (million rubles)

Country	*1971*			*1972*		
	Exports	*Imports*	*Balance*	*Exports*	*Imports*	*Balance*
Algeria	52.6	69.3	-16.7	55.9	58.6	-2.7
Ecuador	0	3.3	-3.3	0.1	2.3	-2.2
Indonesia	10.1	10.1	–	2.6	6.8	-4.2
Iran	139.3	100.1	39.2	95.5	134.0	-38.5
Iraq	99.1	5.5	93.1	90.1	61.6	28.5
Kuwait	17.4	0.7	16.7	14.5	0	14.5
Libya	8.9	0	8.9	8.6	30.0	-21.4
Nigeria	15.7	41.0	-25.3	9.0	19.8	-10.8
Saudi Arabia	5.4	0	5.4	4.5	0	4.5
Venezuela	0	0	–	0.1	4.1	-4.0
Afghanistan	45.3	34.6	10.7	38.1	30.8	7.3

Country	*1973*			*1974*		
	Exports	*Imports*	*Balance*	*Exports*	*Imports*	*Balance*
Algeria	64.7	52.1	12.6	110.3	61.4	48.9
Ecuador	0.2	0.7	-0.5	0.5	4.4	-3.9
Indonesia	2.7	4.2	-1.5	8.0	19.9	-11.9
Iran	137.3	139.6	-2.2	265.8	229.9	35.9
Iraq	141.5	190.6	-49.1	182.3	270.8	-88.5
Kuwait	7.9	0	7.9	4.7	0	4.7
Libya	14.1	30.4	-16.3	28.5	0	28.5
Nigeria	11.0	28.9	-17.9	21.5	70.4	-48.9
Saudi Arabia	2.9	0	2.9	2.8	0	2.8
Venezuela	0.6	0.6	–	0.2	0	0.2
Afghanistan	33.7	35.8	-2.1	61.8	60.6	1.2

Country	*1975*			*1976*		
	Exports	*Imports*	*Balance*	*Exports*	*Imports*	*Balance*
Algeria	112.3	134.7	-22.4	131.4	58.9	72.5
Ecuador	0.6	12.9	-12.3	0.4	7.4	-7.0
Indonesia	7.7	20.9	-13.2	4.4	27.9	-23.5
Iran	281.5	228.2	53.3	217.9	226.7	-8.8
Iraq	270.8	325.4	-54.6	341.6	372.9	-31.3
Kuwait	3.5	0	3.5	10.1	0	10.1
Libya	18.8	0	18.8	16.2	0	16.2
Nigeria	24.3	84.0	-59.7	23.9	26.6	-2.7
Saudi Arabia	5.6	0	5.6	13.2	0	13.2
Venezuela	0.2	0	0.2	0.3	0	0.3
Afghanistan	67.9	64.3	3.6	87.5	66.8	20.7

Source: *Vneshnyaya torgovlya SSSR*, corresponding years

Table 7.2 Oil and Gas in Relation to Soviet Import Trade 1971-1976 (million rubles)

	1971	*1972*	*1973*	*1974*	*1975*	*1976*
Total Soviet Import Trade	11,231.9	13,303.0	15,544.0	18,834.4	26,669.2	28,730.7
Oil and Oil Product Imports	110.5	157.3	273.1	342.5	499.8	497.0
Oil and Oil Products %	1.0	1.2	1.8	1.8	1.9	1.7
Gas Imports	48.5	65.8	84.5	152.1	182.1	176.2
Gas %	0.4	0.5	0.5	0.8	0.7	0.6

Source: *Vneshnyaya torgovlya SSSR*, corresponding years

Soviet Union by the gas suppliers was adjusted in order to maintain the trade balance. In addition, and perhaps of greater significance, is the fact that higher prices for both oil and gas afford the Soviet Union the opportunity of raising gas prices in intra-Comecon trade. Oil prices have risen faster and further than gas prices since October 1973 and the effect has been to stimulate demand for gas. Given the particular problems of the Soviet oil industry and the relative under-development of the gas industry, this induced change assists the Soviet Union in expanding gas consumption and trade. Moreover a rising world price for gas enables the Soviet Union to negotiate better terms with consumers on the expiry of contracts (see table 6.3). The greater foreign exchange burden that would be imposed on Eastern Europe in attempting to increase imports of oil or gas has obliged these countries to provide materials and finance for joint energy projects on Soviet territory that might not otherwise have been undertaken.

Further benefits befall the Soviet Union as a result of higher prices and a decreasing security of supply. There arose, in the aftermath of the Yom Kippur war, a greater interest in the possible joint development of Soviet hydrocarbon resources, particularly from American concerns. From the point of view of the major oil companies, whose operations in the Middle East are gradually coming under the control of the host countries, the Soviet Union's status as an oil supplier may be no less reliable: from the point of view of the American government, however, there may seem to be political disadvantages in assisting a perceivedly hostile superpower in maintaining self-sufficiency in energy. Such a position, as Ian Smart has stated,[65] gives the assisted power substantial flexibility in the conduct of foreign policy in the broad sense.

There is a consistency in the Soviet Union's support of the anti-Israeli and hence anti-American stand of the Arab countries whilst at the same time attempting to negotiate joint developments with American companies and the American government. For example, on the question of the Arab use of the oil embargo as an element of strategy in October 1973, the Soviet Union stated that 'this has been no more than a necessary measure of self-defence. The Arab countries are being condemned because they have turned oil into a political weapon . . . Western sources are trying to ignore the important fact that the Arab countries

Table 7.3 Soviet Imports of Oil and Gas 1971-1976

A. Oil and Refined Products (Volume — thousand tonnes, Unit Price - Rubles per tonne)

Principal Supplier	*1971* Volume	*1971* Unit Price	*1972* Volume	*1972* Unit Price	*1973* Volume	*1973* Unit Price
Iraq	–	–	4,084	14.19	11,010	16.88
Syria	–	–	–	–	247	10.55
Algeria	749	20.65	570	18.29	–	–
Egypt	2,040	6.49	971	9.73	209	6.68
Libya	–	–	1,867	12.30	1,713	17.73
Romania	462	39.18	351.7	40.65	522	32.88

	1974 Volume	*1974* Unit Price	*1975* Volume	*1975* Unit Price	*1976* Volume	*1976* Unit Price
Iraq	3,888	68.44	5,304	60.4	5,821	63.50
Syria	330	48.72	–	–	450	56.48
Algeria	–	–	984	67.20	–	–
Egypt	172	51.27	211	49.54	154	52.90
Libya	–	–	–	–	–	–
Romania	455	33.01	492	56.77	155	66.55

B. Natural Gas (Volume — Million cubic metres, Unit price — Rubles per 1,000 cubic metres)

Supplier	*1971* Volume	*1971* Unit Price	*1972* Volume	*1972* Unit Price	*1973* Volume	*1973* Unit Price
Iran	5,622.6	6.19	8,197.0	6.29	8,679.5	7.80
Afghanistan	2,513.0	5.43	2,849.4	4.99	2,734.9	4.94

	1974 Volume	*1974* Unit Price	*1975* Volume	*1975* Unit Price	*1976* Volume	*1976* Unit Price
Iran	9,094	14.57	9,559	15.33	9,280	15.57
Afghanistan	2,847	6.84	2,853	12.43	2,500	12.61

Source: *Vneshnyaya torgovlya SSSR*, corresponding years

have the full sovereign right to do as they wish with their national resources and to use them in their national interests'.[66] On the day the foregoing statement was issued the view was expressed that 'the energy crisis can be overcome . . . It depends on broad economic cooperation between countries advantageous to all and with discrimination against none . . . There is at times an obsolete approach to international trade in America's policy . . .'.[67] The Soviet Union did not let its commitment to the Arab cause interfere with other negotiations. Again as Ian Smart has pointed out,[68] discussions took place in October 1973 between the US and the Soviet Union on possible joint development of Siberian gas on the day after the Yom Kippur War broke out, in which both countries were heavily involved as arms suppliers.

However, in dealings with the Arab countries the Soviet Union did have to tread warily. It was possible that allegation might be made that they were taking, or intended to take, advantage of the production cutbacks and embargo to expand sales of Soviet oil and refined products in Arab markets, primarily the Netherlands.[69] Examination of the data shows that whereas in 1973 and 1974 deliveries grew at a rate which would scarcely suggest that the Soviet Union was breaking the Arab embargo or eroding their market share by selling at below the world price, the hard currency earnings from such trade contributed to a substantial improvements in the Soviet trade accounts in those years.[70] Since 1973, the Soviet Union has had a larger share of Western Europe's petroleum market. This has been due to depressed consumption in the West rather than to aggressive marketing. Indeed, on occasions Western companies have decided to suspend the purchase of Soviet oil on the grounds that the price asked was higher than could be obtained elsewhere.[71] It seems reasonable to suppose from the increase in deliveries to the Netherlands in 1973, compared with 1972, and from the unit price charged, that the Soviet Union sought to redistribute her available export quota to take advantage of the spot prices available on the Rotterdam market. However, the volume involved did not cause any divergence from the Arabs' intended policy. Soviet export policy in Western Europe at this time was that of riding with the tide. This was shaped by three considerations, the degree of immediate flexibility in the domestic energy balance, the relevance of the Soviet Union's historical commitment to the Arab cause and the changed role of OPEC within the world market.

Taking the energy sector as a whole the view that the Soviet Union has an 'energy crisis' seems somewhat inaccurate. This hinges on the definition of 'crisis'. It is true to say that the Soviet Union has been largely immune to the economic problems that, in the West, have been caused partly by deteriorating relationships between oil producers and consumers and the limiting influence of energy dependence. In the sense that major decisions can be taken by Soviet energy planners largely, though not entirely, without reference to a possible non-indigenous supply, the Soviet Union does not face the same 'energy crisis' as does Japan or the US. Robert Campbell pointed out that the substantial expansion of the role of oil and gas from the early sixties to around 1970 was facilitated by the relative simplicity of production and transportation, whereas in the seventies further expansion demanded the most advanced technology available.[72]

The change of emphasis in Soviet energy policy towards the rehabilitation of coal was mentioned in a speech to the Soviet Academy of Sciences in late November 1974 by V I Kirillin.[73] He stressed the abundance of coal reserves compared with those of oil and gas and outlined the need to reverse the trend towards use of oil and gas in favour of coal, adding that this policy should be undertaken immediately.[74] Despite Soviet statements that there are domestic reserves of all types of conventional fuel, it is now acknowledged that the energy balance is tight (napryazhennyi).[75] However, the essence of the issue is that the Soviet Union does have alternatives to dependence on outside supply which can be put into effect at apparently tolerable cost. This is in contrast to the US, whose 'Project Independence' has already floundered.[76]

The Soviet Union has a long-standing commitment to the Arab cause dating from the early fifties, well before the formation of OPEC. Moreover the nature of superpower involvement in the Middle East has changed, it no longer manifests itself as the active participant in open hostility. Ian Smart has pointed out that the emergence of the Soviet Union as a champion of the Arab cause came about in the early fifties, in the aftermath of the failed Soviet attempt at sponsorship of Israel and the deterioration of relations between Egypt and the West caused by the overthrow of the monarchy. He argued, that as a result of the 'Eisenhower doctrine', launched in 1957, which promised American help to any Middle East nation 'threatened' by international Communism, the Soviet Union took her Middle East role very seriously. By the time of the 1967 six-day war the area was polarized along the lines of an open superpower conflict.[77]

Between 1967 and the full-scale hostilities of 1973, a number of significant events changed the course of international relations, as reflected in policy towards the Middle East. The power of OPEC had increased as had American, Western European and Japanese dependence on Middle East oil. However, the support of each superpower for its respective ally had not waned, despite the major Soviet setback of Sadat's expulsion of the Soviet military and technical advisory team from Egypt in July 1972.[78] The very complexity of these relations helped to prevent an open superpower confrontation and catalyzed negotiations for ending the war itself. Smart described the changed dimension of policy as 'the perennial competition for influence . . . through the provision of military, political and economic support to client states, even if the objective now seems to be more the preservation than the extension of positions already gained'.[79] Indeed the relatively unchanging territorial domain of the major protagonists in the 1967 and 1973 Middle East wars, lends support to Smart's analysis of policy. However, the important issue is that the nature of the forces that determine the equilibrium has altered.

The principal force is now that of the changed power of OPEC and its consequent enhancement of Arab influence. The benefits gained by the Soviet Union from OPEC's production and pricing policy have come about not as a result of Soviet influence within or upon the executive of OPEC. For the reasons outlined, OPEC's policy could have been put into effect in spite of Soviet disapproval. Moreover, it seems hardly likely in the broader context of international political relations, that the Soviet Union would have risked the consequences of withdrawing military support from the Arab states because of its disapproval of OPEC's oil policy. Though Arab and Soviet views might coalesce on the subject of the legality of Israel, the Arab states, many of them members of OPEC, of OAPEC or of both, are likely to be wary of becoming dependent on the Soviet Union despite the communality of interest on this point. However, they are free to exploit it. As Peter Odell has pointed out,[80] the Arab states, acting through OPEC, might well see the undermining of world economic systems, capitalist or communist, as the prerequisite of a more appropriate economic and social order. Accordingly, now that the degree of influence conferred by independence is appreciated, it is not likely to be abandoned to disguised colonialism.

Robert Campbell has stated that 'Soviet production cost is far above Middle Eastern cost and it would be advantageous if the USSR could meet its needs from the Middle East rather than from domestic production', and that 'the resources the USSR is spending in expanding its own oil and gas industry, if invested in the Middle East, would give a much bigger payoff'.[81] This depends on the acceptability of such investment to the host country. Relating investment cost to comparative production cost the major oil companies would likewise derive a bigger payoff from increasing investment in the Middle East, rather than in Alaska or the North Sea, or, for that matter, in West Siberia. The overriding issue is whether OPEC would countenance greater development than at present and the indications are that it will not. The common objective of arresting the rate of depletion of reserves was a major factor in consolidating the strategy of OPEC, and there is no evidence in the aftermath of their success that suggests a change in policy.

From study of this aspect of Soviet energy policy it is difficult to support the view that Middle East oil and gas have been the focus of Soviet interest in the area or that participation in the world market, whether as exporter or importer, has had any influence on broader policy questions. However, detectable progress has been made by the Soviet Union in developing what Landis termed the 'strategic middleman' role. This is particularly well illustrated in a recent tripartite agreement between a Western European consortium, consisting of West German, French and Austrian companies, the Soviet Union and Iran, for the supply of Iranian gas to Europe via the Soviet Union. Under the terms of the agreement, signed in Tehran on 30 November 1975, deliveries of 13.4 billion cubic metres into the Soviet Union's Central Asian gathering system will commence in 1981/1982. The Soviet Union will retain 2.4 billion cubic metres for domestic use and will substitute 11 billion from Tyumen' or Orenburg via the 'Bratstvo' pipeline, of which West Germany will receive 5.5 billion, France 3.66 billion and Austria 1.84 billion.[82] It should be noted that the Soviet Union merely gains a transit fee, expressed in the value of the gas retained for domestic consumption. Payment for the bulk of the gas is in the form of Western equipment supplied directly to Iran from the receiving countries, and negotiated independently of the Soviet Union. The further benefit to the Soviet Union is that this arrangement contributes to the development of the Western European gas market and, for reasons outlined, gas is becoming the preferred Soviet export fuel.

This type of arrangement helps to dispel the view held in a number of quarters that the objective of Soviet interest in Middle East hydrocarbon developments is to weaken the West.[83] On balance, the Soviet Union needs a relatively strong US and Western Europe as willing trading partners. It needs them particularly as suppliers of technological goods and as absorbers of raw materials, including oil and gas, which the Soviet Union can most readily export. This does not overlook the fact that industrial depression in Western Europe, exacerbated by the increased cost of importing oil, has caused many producers to look to Comecon as an alternative to declining domestic markets, given overcapacity in many manufacturing industries.

The policy of the Soviet Union towards Middle East hydrocarbon developments has been characterized by opportunism based on a sound economic rationale. This does not discount the gains that could be described as 'political'. The point is that the Soviet Union does not appear to have sustained economic loss in matters relating to hydrocarbon trade for 'political' gain. Galia Golan has been at pains to point out that OPEC's policy, though it suits the Soviet Union, cannot be viewed as a consequence of Soviet policy, nor in itself as the means by which any Soviet design for strategic dominance in the area could be assisted.[84]

The Soviet Union and Norway

The course of Soviet negotiations with Norway, a major oil producer and the only country operating in the North Sea to enjoy a current surplus of energy, has attracted little attention. Analysis has been confined for the most part to discussion of the periodic wranglings between the Soviet Union and Norway over the demarcation of areas of sovereignty in the Barents Sea, and to the existence or non-existence of an individual continental shelf of the Spitzbergen islands.[85] These issues are, however, of less significance than the fact that as a growing exporter of crude oil, Norway is emerging as a competitor for the Western European market. Moreover Norway has so far declined membership of the International Energy Agency and is following an independent line in respect of the rate of development of her oil and gas resources.

In February 1975, it was estimated that Norway had no need of a production level in excess of 50 million tonnes of oil per year and 40 million tonnes of oil equivalent in the form of gas. This was due to the availability of hydroelectricity, the capacity of the economy to absorb estimated incomes from exports of oil and gas, and the Norwegian government's objective of maintaining a diversified pattern of economic growth, to which income from oil and gas could be expected to contribute for a long time.[86]

In an analysis written in July 1975 Odell suggested that in view of the tensions which existed between Norway and the remaining hydrocarbon-deficient countries of Western Europe, there is 'the basis for a Soviet offer to Norway whereby favourable consideration of its claim in the Barents Sea would be extended in return for acceptance of Soviet protection for Norwegian resources against all outside powers'.[87] Such a policy would, he argued, enable the Soviet Union to use Norwegian oil, produced by Norway in the Barents Sea and made available on Soviet account, to be delivered to established Soviet markets in Western Europe. This would alleviate the economic and logistic difficulties of direct supply.[88]

Odell extended this analysis to suggest that Soviet interest in Norwegian oil and gas formed part of an 'expansionist effort to secure the adhesion of Germany and Scandinavia to her system'.[89] However, this analysis overlooked the fact that despite its non-membership of the IEA, Norway is committed to NATO. Moreover it seems highly unlikely that Norway would be prepared to assist the Soviet Union's further penetration of the Western European market, other than on terms which were economically attractive and which left Norway in control of resource depletion and the total rate of increase of exports.

The Soviet Union and Other Hydrocarbon Producers

In the aftermath of the Yom Kippur War the position of Libya in relation to the Soviet Union and Eastern Europe began to attract greater attention. The independent line taken by Gadafi in imposing the oil price rise of 1970, outside the framework negotiated between OPEC and the major oil companies, and in nationalizing the operations of the British Petroleum Company in November 1971 had met with Soviet approval.[90] The willingness of the Soviet Union to tolerate Gadafi's changeable and unpredictable line in dealings with Moscow stemmed from two possible considerations. In the first instance the possibility of building up an alternative military-economic base in the Southern Mediterranean seemed attractive to Moscow given its deteriorating relations with Egypt after 1972, and the impact of Soviet moves towards some sort of alignment with Libya were enhanced by the detectable strains between Gadafi and Sadat. In the second instance the nature and location of Libyan oil made it a particularly attractive prospect for the Soviet Union.[91] In January and February 1974 Eastern European countries made spot purchases of Libyan oil, and Romania negotiated a supply contract for 3 million tonnes of Libyan oil per year to 1977.[92] There is a more important aspect to the Soviet Union's having access to Libyan oil, namely that it can be used to supply the established market for Soviet oil in Italy. Given the increasing difficulty and cost of supplying the Italian market with indigenously produced oil, Libyan oil could be delivered on Soviet account in return for Soviet arms and the technical expertise required to run production facilities after the BP withdrawal of their staff following Gadafi's nationalization of the BP affiliate.[93] Such supply would decrease by a corresponding amount the loading of expensive West Siberian oil. Though it is not possible to prove this hypothesis conclusively from statistics, the evidence is that direct deliveries of crude oil and refined products from the Soviet Union to Italy have shown a decline in the seventies. Direct Soviet liftings of oil from Libya are recorded only for 1972 and 1973. It seems reasonable to suppose that further oil as repayment for arms supplies was made available and delivered by chartered tanker to Italy. In 1976, upon the possible expiry of the Libyan-Soviet contract, deliveries of indigenous oil were increased. This was reflected in the increase in recorded direct deliveries from the Soviet Union to Italy in 1976 compared with 1975.[94]

A similar opportunity exists in the relations between the Soviet Union and American oil producers, principally Venezuela, Mexico and Ecuador. In return for Soviet aid these countries could offer negotiated quantities of crude oil to be delivered on Soviet account to the existing Cuban market (the cost of direct supply to this market is rising) or to the US as part of the balancing element of a possible 'grain for oil' agreement. A strong reason for US unwillingness to finance extensive Soviet oil developments may well be the fear that the supply quantities negotiated will be 'switched' by the Soviet Union to deliveries from alternative sources at their disposal, in which event the US would have financed the maintenance of Soviet self-sufficiency.

The development of relations between Eastern Europe and non-Soviet oil producers dates from the mid-sixties. In December 1966 it was stressed that there should be greater interest on the part of Eastern Europe in negotiating supplies of oil from the Middle East.[95] In the course of the late sixties each of the Eastern European countries did enter into supply contracts with OPEC producers. Romania signed a ten-year agreement with the Iranian National Oil

Company which provided for Romanian purchase of 1,000 million dollars' worth of crude oil in return for industrial equipment, primarily for the development of the petrochemical industry. Czechoslovakia concluded similar agreements with Iran, involving the supply of general industrial machinery against the purchase of oil over a nine-year period, and with Iraq involving the transfer of refinery and pipeline know-how and equipment. Hungary likewise negotiated an agreement with Iraq for the supply of oil against deliveries of gas- and oilfield equipment, East Germany agreed on direct purchases of Iraqi oil, and Bulgaria agreed to buy direct from Algeria.[96]

References

1. I Smart, 'Oil, The Superpowers and the Middle East', *International Affairs*, January 1977, pp 17-18.

2. H Adomeit, 'Soviet Policy in the Middle East: Problems of Analysis', *Soviet Studies*, April 1975, p 304.

3. *ibid*, p 302.

4. *ibid*, p 295.

5. A Nove, 'On Soviet Policy and Intentions', *Survey*, Vol 22, No 314, Summer/Autumn, 1976, p 112, highlights the fact that Soviet reporting of its own foreign policy has not only a communicative function but also one of reinforcing official doctrine.

6. F D Kohler, L Gouré, M L Harvey, *The Soviet Union and the October 1973 Middle East War: The Implications for Detente*, Washington DC, University of Miami Center for Advanced International Studies, 1974, p 121.

7. J W McKie, in *Daedalus*, Vol 104, No 4, (Fall 1975), p 84.

8. J C Campbell, 'The Energy Crisis and US Policy in the Middle East', in J S Szyliowicz, B E O'Neill (eds), *The Energy Crisis and US Foreign Policy*, New York, Praeger, 1974, p 111.

9. Douglas Evans, *The Politics of Energy: The Emergence of the Superstates*, London, Macmillan, 1976, p 24.

10. J S Nye, 'Multinational Corporations in World Politics', in E W Erickson, L Waverman (eds), *The Energy Question: An International Failure of Policy*, Toronto, University of Toronto Press, 1974, Vol 1, p 213.

11. W Laqueur, *The Struggle for the Middle East: The Soviet Union and the Middle East 1958-68*, Harmondsworth, Penguin, 1972, p 119.

12. For example, governmental systems vary from the parliamentary in Kuwait to conservative autocratic monarchies such as Iran and Saudi Arabia and to revolutionary military governments such as Algeria and Libya. There have been inter-Arab disputes over a number of territorial questions. See Z Mikdashi, 'The OPEC Process', *Daedalus*, Vol 104, No 4, (Fall 1975), p 209.

13. During the 1973 war with Israel 'the Arab oil-exporting countries, although geographically distant from the front line, could not wait in the sidelines and see their Arab compatriots being deprived of their lands . . . The Arab oil-exporting countries will not hesitate to use all the weapons they have and will continue to give every material and moral support . . . in their fight for the restoration of Right and Justice'. M S Al-Otaiba, *OPEC and the Petroleum Industry* (preface to the English language edition), London, Croom Helm, 1975, p vi.

14. Mikdashi, *op cit*, p 204. Though of course the major producers, Saudi Arabia and Kuwait, were members of OAPEC, the full cooperation of OPEC was necessary in order to prevent a recurrence of the failed boycott of 1967.

15. For example, the Kuwait Oil Company, jointly owned by BP and Gulf, each having an equal share of the equity, was due to cede its concession rights to the Kuwaiti government in 2025. By March 1975 it was fully nationalized. *Petroleum Economist*, April 1975, pp 124-125.

16. Saudi Arabia insisted on a 5 per cent rise for Arab light ('marker') oil, supported by the UAE. Other producers opted for a 10 per cent rise effective from 1 January 1977 and a further rise of 5 per cent effective from 1 July 1977. This was the first split in OPEC on the issue of pricing. *Petroleum Economist*, January 1977, pp 2-3.

17. Laqueur, *op cit*, p 133.

18. *ibid*, p 135.

19. S Wasowski, 'The Fuel Situation in Eastern Europe', *Soviet Studies*, July 1969, p 50.

20. Laqueur, *op cit*, p 132.

21. *ibid*, pp 135-136.

22. L Z Zevin, in E Kamenov, G Prokhorov (eds), *Mirovoi sotsializm i razvivayushchiesya strany*, Moscow, Mysl', 1969, p 22. Also Zevin, *Novye tendentsii v ekonomicheskom sotrudnichestve sotsialisticheskikh i razvivayushchikhsya stran*, Moscow, Nauka, 1970, p 123.

23. N P Shmelev, *Problemy ekonomicheskogo rosta razvivayushchikhsya stran*, Moscow, Nauka, 1970, p 222.

24. Zevin, *Novye tendentsii . . .*, (1970), p 127.

25. Zevin, in Kamenov, Prokhorov, *op cit*, p 24.

26. W Gumpel, 'Sowjetunion: Erdöl und Nahostpolitik', *Aussenpolitik*, 11/1971, pp 670-681; Sabine Baufeldt, 'Die künftige Erdöllücke im RGW vor dem Hintergrund des sowjetischen Engagements in Nah-Mittel-Ost', *Osteuropa Wirtschaft*, June 1973, pp 35-54.

27. Gumpel, *op cit*, p 681.

28. Baufeldt, *op cit*, p 47.

29. *ibid*, pp 48-49.

30. *ibid*, pp 51-52.

31. *ibid*, pp 51, 53.

32. *ibid*, pp 53-54.

33. Robert E Hunter, 'The Soviet Dilemma in the Middle East: Part 2, Oil and the Persian Gulf', *Adelphi Paper No 69*, London, International Institute for Strategic Studies, October 1969.

34. *ibid*, p 3.

35. *ibid*, p 7.

36. *ibid*, p 9.

37. *ibid*, p 17.

38. Abraham S Becker, 'Oil and the Persian Gulf in Soviet Policy in the Seventies', in M Confino, S Shamir (eds), *The USSR and the Middle East*, Jerusalem, Israel Universities' Press, 1973, pp 173-214.

39. *ibid*, pp 186-187.

40. *ibid*, p 175.

41. *ibid*, p 184.

42. *ibid*, p 185.

43. *Ekonomicheskaya gazeta*, 22/1969, pp 12-13; *Izvestiya*, 9 December 1970, p 3, 19 December 1972, p 2.

44. Becker in Confino, Shamir (eds), *op cit*, p 190.

45. *ibid*, p 195.

46. Abraham S Becker, 'Soviet-American Relations after the Energy Crisis', in Szyliowicz, O'Neill (eds), *op cit*, pp 159-182.

47. *ibid*, p 177.

48. *ibid*, pp 177-178.

49. Lincoln Landis, *Politics and Oil: Moscow in the Middle East*, New York, Dunellen, 1973.

50. *ibid*, pp 95-96.

51. *ibid*, p 101.

52. *ibid*, p 121.

53. Galia Golan, *Yom Kippur and After: The Soviet Union and the Middle East Crisis*, London, CUP, 1977, p 13.

54. *ibid*, p 15.

55. *ibid*, p 243.

56. Arthur Jay Klinghoffer, *The Soviet Union and International Oil Politics*, New York, Columbia University Press, 1977, pp 118-122.

57. Adomeit, *op cit*, pp 294-295.

58. *Middle East Economic Digest*, 11 July 1975, p 26.

59. *The Middle East*, 10 July 1975, p 24.

60. G Ofer, 'The Economic Burden of Soviet Involvement in the Middle East', *Soviet Studies*, January 1973, p 333.

61. Kohler, Gouré, Harvey, *op cit*, pp 74-75.

62. In comparing the unit prices of Iranian and Afghan gas to the Soviet Union it must be borne in mind that Iran constructed and paid in hard currency for the pipeline from its gas-fields to the Soviet border, whereas in the Afghan case the Soviet Union undertook to finance and build the pipeline.

63. *Financial Times*, 11 July 1974, p 8.

64. *ibid*, 19 August 1974, p 5.

65. Smart, *op cit*, pp 30-31.

66. *Moscow Radio* (in Arabic), 16 November 1973.

67. *ibid* (in English), 16 November 1973.

68. I Smart, 'The Superpowers and the Middle East', *The World Today*, January 1974, p 14.

69. The strategy adopted by the Soviet Union for avoiding such charges is outlined in detail in Golan, *op cit*, pp 196-202.

70. It should be noted that Soviet statements on export policy made in 1974 were somewhat contradictory. The Soviet Oil Minister, V D Shashin, is reported to have said that the Soviet Union was unlikely to increase oil exports significantly (*Financial Times*, 29 May 1974, p 6). However, it is subsequently reported that TASS refuted this statement (*Financial Times*, 7 June 1974, p 4).

71. For example, the West German oil and petrochemical company VEBA suspended deliveries from the Soviet Union in March 1974 when the delivered price was stated to be the equivalent of $17 per barrel, *Financial Times*, 14 March 1974, p 24.

72. R W Campbell, *Trends in the Soviet Oil and Gas Industry*, Baltimore, John Hopkins Press, 1976, pp 86-87.

73. V I Kirillin, 'Energetika: problemy i perspektivy', *Kommunist*, 1/1975, pp 43-51.

74. *ibid*, p 46.

75. P Neporozhnyi, 'Perspektivy Sovetskoi energetiki', *Planovoe khozyaistvo*, 8/1975, p 44.

76. R El Mallakh points to the contradictions in 'Project Independence', in 'American-Arab Relations: Conflict or Cooperation?', *Energy Policy*, September 1975, p 170.

77. Smart, 'The Superpowers . . .', *op cit*, pp 7-8.

78. Galia Golan points out that in the build-up to the Yom Kippur War the sole issue on which the Soviet Union and Egypt were in agreement was that of the possible use of the 'oil weapon', *op cit*, pp 55-56.

79. Smart, 'The Superpowers . . .', *op cit*, pp 14-15.

80. P R Odell, *The Western European Energy Economy*, Leiden, Stenfert Kroese, 1976, p 27.

81. Campbell, *op cit*, p 81.

82. *Financial Times*, 1 December 1975, p 5; *Petroleum Economist*, January 1976, p 5.

83. A view expressed by J A Berry in 'Oil and Soviet Policy in the Middle East', *The Middle East Journal*, Vol 26, No 9, Spring 1972, p 149.

84. Golan, *op cit*, p 250.

85. Background to the debate is given in Jeremy Russell, *Energy as a Factor in Soviet Foreign Policy*, Farnborough, Saxon House, 1976, pp 185-186, and by P Hill in *The Times*, 16 November 1976, p 16. Briefly, dividing the Barents Sea under the terms of the 1958 Continental Shelf Act by a median line between Norway and the Soviet Union would give Norway the greater share. The Soviet Union favours the system adopted in the Antarctic, namely sectoral extension radiating from the Pole, which would give the Soviet Union the greater share. In the case of Spitzbergen, the 1920 Agreement gives the 41 signatory countries the right to prospect for minerals provided they adhere to Norwegian law. If Spitzbergen is deemed to have its own continental shelf these countries have the right to drill offshore. Norway has argued that Spitzbergen is an extension of its own continental shelf, in which case the Norwegian government would have the right to control exploration.

86. The background to Norwegian oil policy in the seventies given by B S Aamo in 'Norwegian Oil Policy: Basic Objectives', in M Saeter, I Smart (eds), *The Political Implications of North Sea Oil and Gas*, Guildford, IPC, 1975, p 88.

87. P R Odell, 'The World of Oil Power in 1975', *The World Today*, July 1975, p 278.

88. *ibid*.

89. *ibid*, p 281.

90. In 1970 the Soviet Union delivered 200 tanks, field guns and amphibious vehicles to Libya: in April 1972 Gadafi withdrew the Libyan ambassador from Baghdad as a protest against the signing of a 15-year treaty of cooperation between Iraq and the Soviet Union. However, after nationalizing BP's Libyan assets in November 1971 and finding that existing customers were unwilling to purchase oil produced from nationalized fields – especially since BP threatened legal action against anyone doing so – Gadafi turned again to Moscow in May 1972. After reconciliation, Gadafi negotiated an agreement whereby Soviet technicians operated the oilfields in question and the Soviet Union accepted a share of the oil produced. See *Petroleum Press Service*, February 1972, pp 64-65; R M Burrell, 'The Soviet-Libyan Arms Deal', *Soviet Analyst*, 5 June 1975, pp 2-3.

91. Libyan oil is of the 'light' variety, ie suited to the maximal output of non-substitutable products and commanding a premium in Western European markets at that time.

92. Golan, *op cit*, p 199.

93. Burrell, *op cit*, p 3.

94. Analysis of this is further complicated by the decline in total demand caused by the OPEC price rises, reflected in falling total imports.

95. 'Close attention is now being given in almost all the Comecon countries to the necessity of . . . improving the fuel and energy balance . . . In spite of the import of large quantities of oil and gas from the Soviet Union the intensified development of the Czechoslovak economy calls also for the import of oil from the developing countries.' *Rudé právo*, 22 December 1966, p 4.

96. The first agreements between Comecon member countries and OPEC producers are recorded in *Mizan*, August 1971, p 31.

Chapter 8

Summary Analysis and Prospects to 1985

Though insulated from the major upheaval experienced by the world's major oil consumers during the seventies, the Comecon bloc has begun to encounter a number of problems. These have slowed the growth rate of oil within the Soviet energy balance, altering the economic relationship of refined products to alternative forms of energy and feedstock and necessitating readjustment of energy policy in the bloc. During the ninth Five-Year Plan the Soviet Union's energy consumption reached 4 billion tonnes of standard fuel, of which 2.6 billion were used in electricity generation and industrial steam raising. Energy exports totalled 1 billion tonnes of standard fuel.[1] Soviet energy planners have begun to acknowledge openly that the role of oil and gas must be reappraised. Current policy seeks to phase out the use of residual fuel oil in favour of Ekibastuz and Kansk-Achinsk coal for power stations in regions to the east of the Urals, and to accelerate the programme of nuclear power station construction in the European zone.[2] It is now admitted in the Soviet Union that failure to prove new oil reserves at an adequate rate during the ninth Plan has caused tensions in energy planning and has brought about the rehabilitation of coal, and the need to accelerate the development of nuclear power.[3]

However, this policy cannot solve entirely the problems faced in the Soviet and Eastern European energy sector to 1980 and beyond, since the only readily substitutable fuel product derived from oil is residual fuel oil. Moreover the process of fuel substitution is conveniently applicable only in respect of new installations. In the absence of substantial administered price advantages for coal, gas or other fuels, reflected in the final energy cost to the consumer, the increasing cost of converting existing plant appears, in the majority of cases, to outweigh the energy cost saving. Though fuel substitution is openly discussed in recent Soviet analyses it seems unlikely that the energy supply position will be markedly eased in the period to 1980.

The sharp decline in Soviet production of oil and gas condensate, considered likely by the Central Intelligence Agency of the US, is thought to be inaccurate. There is comparatively little technical difficulty in producing the desired 1980 level from existing operations, though it must be admitted that this could only be achieved at a cost considerably higher than estimated for the tenth Plan, and with severe long-term loss. It should be stressed that on the basis of information published up to 1977, the oil industry's performance is near target and this seems to indicate a 1980 performance at the lower end of the original Plan range, namely 620 million tonnes. Rates of increase in production are declining:

this has been anticipated by planners. However, if sufficient new reserves are not discovered during the tenth Plan, the point at which Soviet energy planners are faced with the problem of optimizing the energy balance to an oil production level that has reached its peak, may come in the late eighties. Nonetheless, available Soviet material gives no indication that planners and energy analysts foresee this eventuality, in spite of their acknowledgement of the problems faced by the industry.

Given the difficulties of forecasting and planning for the gas industry and its consistent failure to meet targets in the past, it might be thought reasonable to regard the Plan for 1980 as wishful thinking. However, certain features should be stressed. First, in the latter years of the ninth Plan the increases in production were as originally intended, and this has continued in the first two years of the tenth Plan. Second, the major new discoveries are geographically close to one another, thus easing the logistic problems provided that the delivery systems are completed near schedule. Substantial assistance from Eastern Europe has been obtained and labour transferred from other sectors of the economy to bolster the oil and gas industries' labour force. The administrative mechanism of the gas industry has been modified, with the objective of improving communications by reducing the number of administrative levels.[4]

A leading Soviet specialist on the gas industry admitted that prior to 1972 annual plans were over-ambitious in relation to resources available. This was conditioned by the fact that initial discoveries were particularly large. He argued that the period since 1972 has been characterized by more realistic annual plans, though admittedly showing a considerably reduced rate of growth for the ninth Plan as a whole.[5] The gas industry has recorded an auspicious two years at the outset of the tenth Plan. In terms of volume the industry needs to record an annual increase in production of 28.5 billion cubic metres in order to reach the 1980 target of 435 billion cubic metres. These increases are only slightly above those recorded during the period which Smirnov regarded as having been more realistically planned.

Both the oil and gas industries of the Soviet Union fulfilled the annual production Plan in 1976. In 1977 the gas industry overfulfilled, whilst the oil industry recorded a shortfall of 5 million tonnes. The coal industry likewise underfulfilled its annual target in 1977. The aggregate Plan for the three principal fuel industries, when converted into units of standard fuel, showed a 99.4 per cent fulfilment for the year. Considering the declining rates of growth in major sectors of the economy, and the underfulfilment of the 1977 annual Plan by a number of energy-intensive industries, this figure might well indicate that the energy intensity of the economy as a whole is not declining at the rate desired by Soviet planners.

The Soviet Union has undertaken to supply fellow-members of Comecon in the course of the tenth Plan with 364 million tonnes of oil and 90 billion cubic metres of natural gas.[6] It should be noted that the former figure relates to crude oil only and that both figures include deliveries to Cuba and Mongolia but exclude those to Yugoslavia, an associate member of Comecon. Comparison of these figures with the corresponding commitment and deliveries during the ninth Plan illustrates the importance now attached to natural gas as the growing hydrocarbon fuel in Eastern Europe.

A range of estimates of the 1980 Soviet energy balance is presented in table 8.1. The major factors determining the final balance are the planned decrease in energy intensity in the economy as a whole and in particular in electricity generation. It is felt that there is scope for further improvement in this area. The rate of increase in the automobile park has begun to decline and the conversion of the railroads to oil power is well advanced. Hence the Soviet estimate that the share of oil in the domestic energy balance is not now expected to rise (see p 134) and plans for refinery expansion indicate that a greater share of oil production can be directed to the petrochemical industry and to export. If therefore the 1980 share of the energy balance held by oil is put at the 1975 level, namely 42 per cent, and it is assumed, that for reasons of energy efficiency fuels such as shale and peat will not show an increased share, and that nuclear power is unlikely to make an impact by 1980, then approximately 50 per cent of the Soviet energy balance will be met by coal and gas. Given the greater exportability of gas in comparison with coal, it is likely that Soviet planners would favour reliance on coal for domestic consumption. It is therefore concluded that coal is likely to account for 27 per cent and natural gas for 23 per cent of the 1980 Soviet energy balance.

Corresponding estimates for the energy balance of the Eastern European member countries are given in table 8.2. A remarkable consistency can be discerned in the estimates, the noteworthy exception being Russell's estimate for Romania, which assigns a particularly high share to coal. Since Russell did not state the sources for his table or the basis for his calculations it is not possible to account for this discrepancy: however, it is important to note that it is inconsistent with Western and Eastern estimates. The estimate in this volume takes into

Table 8.1 Estimates of Soviet Energy Balance in 1980 (%)

Estimate	*Coal*	*Oil*	*Natural Gas*	*Other*	*Total*
Bethkenhagen	31.4	43.1	25.5	–	100
Russell	28.5	38.7	24.7	8.1	100
Hanson	26.0	42.0	23.0	10.0	100
Scanlan	32.1	37.9	28.4	1.6	100
Author	27.0	42.0	23.0	8.0	100

Sources: (i) J Bethkenhagen, *Bedeutung und Möglichkeiten des Ost-West-Handels mit Energierohstoffen*, Deutsches Institut für Wirtschaftsforschung, Sonderheft 104, Berlin, Duncker & Humblot, 1975, p 148. Bethkenhagen simply divides his classification of 'primary energy' between coal, oil and gas, ignoring the contribution of the minor fuels and hydroelectricity; (ii) J Russell, *Energy as a Factor in Soviet Foreign Policy*, Farnborough, Saxon House, 1976, p 33; (iii) P Hanson, 'The Soviet Energy Balance', *Nature*, 5 May 1976, p 3 (total exceeds 100 due to rounding); (iv) A F G Scanlan, 'The Energy Balance of the Comecon Countries' in *Exploitation of Siberia's Natural Resources*, Brussels, NATO, 1974, p 97. Note that the category 'other' includes only hydroelectric and nuclear power and that the figure given here for coal includes other solid fuel. For the purpose of this study the important fact is that the categories 'oil' and 'natural gas' should be comparable.

Table 8.2 Estimates of the Eastern European Energy Balance in 1980 (%)

Country	*Fuel*	*Bethkenhagen*	*Russell*	*Bednarz*	*Maksakovskii*	*Author*
Bulgaria	Coal	31.4	31.6	25.0	30.0	29.6
	Oil	56.0	50.0	43.0	48.4	49.0
	Natural Gas	12.6	18.4	23.0	17.4	16.4
	Other	–	–	9.0	4.2	5.0
Hungary	Coal	27.1	25.7	27.9	28.0	21.0
	Oil	50.1	48.6	42.6	39.6	52.4
	Natural Gas	22.8	25.7	28.3	21.3	18.2
	Other	–	–	1.2	11.1	8.4
GDR	Coal	61.2	62.5	68.5	61.4	61.4
	Oil	29.2	22.9	23.0	22.4	22.4
	Natural Gas	9.6	14.6	6.9	11.5	11.5
	Other	–	–	1.6	4.7	4.7
Poland	Coal	67.7	64.0	70.5	66.6-70.3	70.5
	Oil	21.2	20.0	19.0	18.0-19.0	19.0
	Natural Gas	11.1	16.0	8.5	8.4-8.5	8.5
	Other	–	–	2.0	2.2	2.0
Romania	Coal	22.5	39.3	25.0	24-25	23.5
	Oil	37.1	38.7	30.0	64- }	34.3
	Natural Gas	40.4	22.0	35.0	65 }	36.3
	Other	–	–	10.0	10-12	5.9
Czechoslovakia	Coal	61.1	62.6	60.0	57.0	65.7
	Oil	30.2	30.0	30.0	29.7	25.1
	Natural Gas	8.6	7.4	6.0	7.5	6.0
	Other	–	–	4.0	5.8	3.2

Sources: (i) J Bethkenhagen, *op cit*, pp 150-155, (As in table 8.1 Bethkenhagen divides his 100 per cent 'primary energy' balance between coal, oil and natural gas); (ii) J Russell, *op cit*, p 119, (Calculated from data expressed in Mtsf. Like Bethkenhagen, Russell divides primary energy between coal, oil and natural gas, but notes that total 1980 estimated demand of 615 Mtsf should have an additional 17 Mtsf (2.6 per cent) from 'minor sources'.); (iii) L Bednarz, 'Problemy naftowe socjalistycznej integracji gospodarczej', *Nafta (Krakow)*, 4/1974, p 532; (iv) V P Maksakovskii, *Toplivnaya promyshlennost' sotsialisticheskikh stran Evropy*, Moscow, Nedra, 1975, pp 41-42; (v) Author's estimates are based on the following sources: T Khristov, 'Novi tendentsii v razvitieto na energetikata baza v Bolgariya', *Geografiya (Sofiya)*, 1970, Vol 20, No 8, p 2; V Bese, 'Hungary's Mineral Oil and Gas Industry', *Marketing in Hungary*, 4/1971, p 8; Bednarz, *op cit*, p 532; I V Herescu, 'Dezvoltarea bazei energetice (1)', *Revista economică*, 28/1976, p 1; *Hospodářské noviny*, 37/1976, p 3.

account the most recent primary source material available and, as outlined in the analysis contained in chapters 3 and 4, evidences the rehabilitation of coal and the emergence of natural gas, given restrictions in the availability of oil. It is likely that energy consumption in 1980 will be at the lower end of the ranges indicated in table 4.1. It is certain that attempts will be made to consume less energy, particularly oil, which could even lead to a lower share for this fuel than indicated in table 8.2.

Apart from restrictions in the form of the delivery quota stipulated by the Soviet Union and the difficulties of purchasing significant quantities of oil from OPEC, the effect of the new pricing formula for Soviet oil is likely to bring about a measure of oil conservation. The cost of a given quantity of oil imported by an Eastern European country from the Soviet Union in 1977 will be some 40 per cent above that for 1975. Prices for 1978 and beyond depend on the policy of OPEC, and should rise markedly once the pre-Yom Kippur prices are eliminated from the five-year moving average. Even without oil quota restrictions this would tend to increase demand for gas, the price of which has not risen as dramatically as that of oil.

It was estimated in chapter 6 that the Soviet Union is likely to have between 150 and 170 million tonnes of oil and refined products available for export in 1980. The estimate of 1980 energy demand for the Comecon countries given in table 4.1, of the energy balance of the individual countries outlined in table 8.2, of the likely upper limit of 30 million tonnes of oil imported from OPEC producers and of Eastern European production (primarily Romanian) of 18 million tonnes, all imply a requirement of approximately 75 million tonnes from the Soviet Union. This leaves a maximum of 97 million tonnes for export to the rest of the world, and a minimum of 75 million.

Better information is available on Soviet natural gas export commitments than is available for oil. This information is detailed in table 8.3. It was suggested in chapter 6 that Soviet export availability of natural gas was likely to be of the order of 59 billion cubic metres. The possible excess of a few billion cubic metres over commitment would enable the Soviet Union either to enter further export contracts provided that production is on target towards the end of the tenth Plan, or to accelerate the process of substitution of fuel by gas in the domestic energy balance.

The foregoing analysis suggests that the Soviet Union will continue to play the role of a concerned observer in its dealings with members of OPEC. It is likely to use its involvement to enhance opportunities of building up a position in the international oil broking system. In this sense the Soviet Union could become increasingly 'dependent' on the supply of oil from OPEC, should it supply established markets with 'switched' oil.

The prospects for oil and gas production and trade are difficult to assess beyond 1980. Shortage of information is a principal drawback: this shortage is largely due to the existing flexibility in Soviet energy planning and to the growing Soviet difficulty in redirecting energy policy with costs rising in practically all the fuel sectors. There are four major factors that will influence the Soviet development pattern after 1980: the degree of success recorded in joint ventures on Soviet territory, the impact of new technology on exploration and production,

Table 8.3 Soviet Natural Gas Export Commitments 1980 (billion cubic metres)

	Volume
A. Comecon	
Bulgaria	6.3
Hungary	4.0
GDR	5.0
Poland	4.5
Romania	2.8
Czechoslovakia	6.0
B. Comecon Associated	
Yugoslavia	3.0
Sub-Total	31.4
C. Western Europe	
Finland	1.4
Italy	7.0
Austria	2.4
France	4.0
West Germany	8.5
Sub-Total	23.3
Total Export Commitment	54.7

Sources: Russell, *op cit*, p 70; *Petroleum Economist*, May 1977, p 200

the capacity to generate credit in hard currency markets and the involvement of countries with Middle Eastern oil and gas producers.

The American analysts Jack, Lee, and Lent, writing in 1976, identified the need to continue to reduce the rate of increase in oil and gas consumption as the most logical step in Soviet energy policy. They also stated that the Plan for the coal industry through to 1980 could be met without great difficulty.[7] The coal industry's performance in the first two years of the tenth Plan sheds some doubt on this prognosis. Natural gas may have the lowest marginal production cost of the major fuels but it has the highest storage cost. It makes sense to secure further utilization of this fuel in the medium term. The joint Comecon project aimed at rapidly developing the gas and gas condensate reserves of the Orenburg oblast' is perhaps the most important single project currently in progress in the Soviet Union. Soviet planners admit that some large fields of West Siberia, currently producing at what is very likely their peak, will show a decline in production between 1981 and 1985. They are, however, at pains to point out that of over 150 fields in West Siberia that have been fully explored and judged to be economic only 25 are on stream in 1978. This fact has been overlooked by most Western analysts, including the American Central Intelligence Agency.

It seems probable that projects of the Orenburg type, which involve Eastern European countries supplying human and physical resources to the Soviet Union, will increase in the medium-term and become a permanent feature of the development of Soviet fuel resources. Some measure of the confidence felt by Soviet planners as to their ability to solve the type of problem that the CIA regards as insoluble can be gauged from their pursuing the negotiation of export supply contracts for oil and gas well into the eighties.

Since 1970 there has been greater urgency on the Soviet part to negotiate the input of advanced Western technology for use in oil and gas production. The acquisition of this technology would certainly enable the Soviet Union to accelerate its exploration program and to extract further quantities of oil and gas from deposits currently judged uneconomic. Offshore operations are the principal area of activity which would benefit from an injection of Western technology. Broadly speaking, the present stage of development of the Soviet Union's offshore hydrocarbon province is parallel to that of Western Europe in the early sixties, and discussions about offshore development based on Western exploration and production technology are in progress.[8] If these discussions prove fruitful, judgements of the medium-term hydrocarbon potential of the Soviet Union will need to be revised upwards.

The Soviet Union's capacity to absorb Western technology depends, inter alia, on its ability to negotiate hard currency loans on world markets. At the 1976 exchange rate, the Soviet debt accumulated in convertible currencies over the five year period 1972-1976 was 11.4 billion dollars of which 6.4 billion dollars was accounted for in the imbalance in visible trade with the United States.[9]

There are a number of reasons why credit and financial questions should not restrict further development of trade in technology despite the Soviet trade and payment deficits. The Soviet record of debt management and repayment is particularly good. Western state-backed credits have the highly desirable effect of maintaining aggregate production in the domestic market at a time of low demand, and consequently act as a production cost deflator. The Eurocurrency market in particular was growing rapidly even before the recession and the inflow of petrodollars in 1974, and continues to do so. Until the West recovers from its economic depression, the steadily expanding Comecon economies present an attractive market. This is particularly true of the Soviet Union, where reserves of gold, gas and oil are strong collateral securities. It is worth noting that there is a cautious attitude towards the question of the debt service ratio in hard currency trade in the directives of the tenth Five-Year Plan.[10] Above all, it should be borne in mind that the total Soviet debt is a good deal smaller than that of a number of developing countries which are less able to repay.

There is some scope for the negotiation of joint projects of the 'Adria' pipeline type, with Middle Eastern oil and gas producers (discussed in Appendix A). Such projects could involve a Soviet commitment to purchase petrochemical feedstock as well as crude oil from the Middle East, since the major producers are seeking to diversify into the petrochemical sector. There is also, currently, a high level of surplus capacity in Western European industry which limits that area as a potential market.

The Soviet Union has undertaken to continue investment in the North Rumailah oilfields of Iraq. Similar opportunities will be available to other Comecon countries in the medium-term, since at the time of writing there is a surplus of oil in the Western European market and the real value of OPEC's income has fallen in recent times. Consequently, marginal supply to Comecon has become attractive to a number of Middle East producers needing to finance further industrialization. The overriding economic consideration for Middle East producers is to avoid committing themselves to supplying Comecon on terms which enable the Soviet Union to make inroads into the Western European market. In the political sphere, as Jeremy Russell has pointed out,[11] the West should be more concerned with finding a communality of interest with Middle East oil producers, than with opposing or neutralizing perceived Soviet advances in the area.

Despite the need for Western countries to diversify not only the structure of their energy balance but also their source of oil supply, the Soviet Union will remain only a marginal supplier. It is misleading to talk in terms of the Soviet Union's 'loss of opportunity' for increasing her hard currency earnings by supplying Comecon rather than the West. On the one hand the greater flexibility in pricing and output strategy enjoyed by OPEC in its major export market would be adequate to counter any Soviet attempt to enlarge sales to the West which could reduce the economic and political influence the Organization now exercises. On the other hand the oil companies, having committed themselves to high-cost exploration in the North Sea and elsewhere, have an interest in the maintenance of a high price and, as far as possible, a balanced supply and demand for oil in Western Europe.

Recognizing the desirability of acquiring Western technology to develop its hydrocarbon resources the Soviet Union was able to support OPEC's politico-economic strategy towards Western companies and governments. The effect of increasing oil prices and the uncertainty of future supplies from OPEC served to make joint projects somewhat more attractive to the major companies. Though it can hardly be said that the Soviet Union catalyzed the price rises and production cutbacks, the country certainly benefited from them. The Soviet Union was able to increase substantially its hard currency earnings through hydrocarbon exports to Western Europe and to negotiate large price increases for exports to fellow members of Comecon. This latter outcome brought Eastern Europe closer to the Soviet Union in that provision of capital, material and human resources for the development of Soviet hydrocarbon reserves has become the optimal solution of Eastern Europe's energy problem in the medium-term. Comecon members are therefore more closely tied together, and also contribute to the maintenance of self-sufficiency in energy, most importantly in hydrocarbons, of the bloc's principal economic and political power.

References

1. M M Brenner, 'Effektivnost' ispol'zovaniya toplivno-energeticheskikh resursov', *Voprosy ekonimiki*, 6/1977, p 38.

2. *ibid*, p 40.

3. V M Gzovskii, 'Perspektivnoe napravlenie razvitiya neftyanoi promyshlennosti SSSR', *Izvestiya AN SSSR Ser. ekon.*, 3/1977, p 73.

4. The reform involved a reduction in the number of decision-making levels from five or six to two or three. This is discussed in some detail in Appendix C of Jeremy Russell, *Energy as a Factor in Soviet Foreign Policy*, Farnborough, Saxon House, 1976, pp 223-226.

5. V A Smirnov, 'Gazovaya promyshlennost', *Ekonomika i organizatsiya promyshlennogo proizvodstva*, 5/1975, p 49.

6. I Ivanov, A Loshchakov, 'Sotrudnichestvo stran SEV i vyravnivanie urovneĭ ikh razvitiya', *Voprosy ekonomiki*, 6/1977, p 12.

7. E E Jack, J R Lee, H H Lent, 'Outlook for Soviet Energy', in *The Soviet Economy in a New Perspective*, Washington DC, US Congress, Joint Economic Committee, 1976, p 472.

8. J D Park, 'Hydrocarbon Policy and Offshore Developments in Comecon', *Offshore Oil & Gas Yearbook 1978/79 UK & Continental Europe*, London, Kogan Page, 1978, pp 193-200.

9. Discussed in detail in M C Kaser, 'American Credits for Soviet Development', *British Journal of International Studies*, 3/1977, esp pp 143-144.

10. This is discussed in detail in R Portes, 'West-East Capital Flows: Dependence, Interdependence and Policy', University of London, Birkbeck College, *Discussion Paper No 50*, April 1977.

11. Russell, *op cit*, p 203.

Appendix A

Joint Ventures in the Development of Soviet Oil and Gas Resources

Information on joint projects in the development of the Soviet oil and gas industries has been incorporated in a number of recent studies. In order to complete this information the current status (1977) of these projects is included here. For a fuller history of the negotiations of joint projects the reader is referred to the following works: Robert E Ebel, *Communist Trade in Oil and Gas*, New York, Praeger, 1970. Chapters 8 (oil) and 9 (gas) outline the course of negotiations commenced in the mid-sixties. Jeremy Russell, *Energy as a Factor in Soviet Foreign Policy*, Farnborough, Saxon House, 1976. Post-1970 developments are outlined in a series of chapters comprising part 2 of this book, giving a country-by-country analysis.

This appendix is based on the author's article in E de Keyser (ed), *The European Offshore Oil and Gas Yearbook 1976/1977*, London, Kogan Page, 1976, updated where necessary to include the most recent information available.

Soviet-American-Japanese Development of East Siberia

This scheme is concerned with the development of the gas reserves of Yakutia. Negotiations involve cooperation between the American Occidental Oil Company, El Paso Natural Gas Company and the engineering company Bechtel, in conjunction with the Japanese companies Tokyo Electric Power and Tokyo Gas. Preliminary inconclusive discussions have been in progress since the mid-sixties, but in the aftermath of 1973-1974 there was a new momentum towards a possible conclusion. Early in 1974 the tripartite talks brought agreement on the basis of the provision of a 3,400 million dollar loan to be contributed in equal amounts by the US and Japan. It was intended that this loan would be used by the Soviet Union for the purchase of equipment to construct a 3,220 kilometre pipeline to the port of Nakhodka, where a liquefaction plant would be sited. Deliveries of 30 billion cubic metres of gas per year would commence in 1980, to be divided equally between the Soviet Union, the US and Japan over a 25-year period. The American negotiators anticipated that El Paso would receive 75 per cent and Occidental 25 per cent of the American allocation, with the provision that Occidental would accept increased quantities if reserves and production proved greater than anticipated.

Soviet-Japanese Development of West Siberia (Tyumen')

Discussions between the Soviet Union and Japan concerning the possibility of joint development of oil and gas in West Siberia commenced in 1966. Originally the Soviet Union was prepared to supply 40 million tonnes of oil per year, but in the light of problems encountered nationally in maintaining oil production this figure was reduced in 1973 to a maximum of 25 million. At the same time the Soviet Union indicated that as a result of Western inflation and changes in exchange rates that worked to the disadvantage of the Soviet Union, considerably higher credits than originally envisaged, would be required. Japan's initial reaction was that these new terms were still of interest: however, in the early part of 1974 Soviet insistence that the oil would be supplied to Japan at the current world price, indicated that the future price would be linked to the Middle East price. This, together with the Soviet Union's expressed wish to use part of the loan to finance the construction of a second Trans-Siberian railway in preference to a pipeline, evoked a negative response from Japan, which was unwilling to become involved in developing the means by which the Soviet Union could improve its military logistic system close to the Chinese border. At the time of writing negotiations are still at stalemate.

Soviet-American Development of North-West Siberian Gas Reserves

This scheme involves a consortium of three American companies, Tenneco Petroleum, Texas Eastern Gas and the engineers Brown and Root. Negotiations have centered on the construction of a pipeline from Urengoi to Murmansk, where liquefaction facilities would be sited. The envisaged cost is of the order of 7.6 billion dollars, of which the Soviet Union agreed to raise 1.5 billion dollars, the rest to be raised by the American consortium. Deliveries of gas would commence in 1980.

Schemes involving American participation experienced a strong setback in December 1974, when Congress decreed that the lending authority of the Export-Import Bank would be limited to a maximum of 25 billion dollars over a four-year period, with a subceiling of 40 million dollars on loans in the energy sector. In addition, it was stipulated that 25 legislative days' notice should be given for energy project proposals involving loans of 25 million dollars and above. Subsequent attempts have been made to raise finance from sources other than the Export-Import Bank: however, there are currently only two projects under discussion, the Yakutia gas project and a proposal to develop offshore oil reserves in Sakhalin. A revised agreement for the former project was signed in April 1976 providing initial credits of some 50 million dollars, of which 25 million dollars would be provided by the Bank of America, 20 million dollars by the Japanese Export-Import Bank and 5 million dollars from private Japanese sources. In the case of Sakhalin an agreement was signed by Japan, the US and the Soviet Union in January 1975 covering oil and gas developments. Initial credits of 600 million dollars were to be provided by the Japanese Export-Import Bank and private industry.

Anglo-Soviet Development of Offshore Oil Reserves (Caspian)

In April 1976 an agreement was reached between Britain and the Soviet Union concerning the provision of British technology, proved in the North Sea, for use in the development of offshore reserves in the Caspian Sea. The companies most closely involved in this project are BP and Highland Fabricators Ltd, the engineering company operated jointly by Wimpey and Brown and Root. The agreement provides for the supply of drilling equipment and expertise.

Smaller Scale Joint Projects

A number of smaller-scale joint projects involving Western countries in the development of oil and gas reserves in Comecon have been initiated. In March 1974, the governments of Sweden and Finland signed a ten-year agreement with the Polish government covering joint exploration of the continental shelf of the Baltic Sea, following a Polish oil strike. The government of West Germany has had preliminary discussions with the Soviet Government on the possibility of joint exploration in the Barents Sea, though it seems more likely that negotiations with the Norwegian government will prove more fruitful. Bulgaria and Romania, recognizing the possibility of using proven offshore technology, have invited interested Western companies for preliminary discussions.

The 'Adria' Pipeline

The history of the joint participation of Hungary, Czechoslovakia and Yugoslavia with Kuwait in the construction of a pipeline from the island of Krk near Rijeka to Zagreb, with branches to Hungary and Czechoslovakia, is given as Appendix E to Russell's *Energy as a Factor in Soviet Foreign Policy*, (pp 228-229). The Hungarian branch is intended to link up with the trans-Comecon 'Druzhba' pipeline at Szazhalombatta for forward transmission to Bratislava; the Yugoslav branch is to extend to the north and north-east of the country to feed refineries at Bosanski Brod, Novi Sad and Pancevo. Russell estimated the total cost of the project at 500 million dollars. The section from Krk to Zagreb was due to be operational by early 1977: however, construction work to expand port facilities only commenced in mid-1977. (See *Petroleum Economist*, July 1977, p 285.)

Appendix B

Utilization of Natural Gas in the Soviet Union, 1971-1975

The information contained in this appendix is derived from a single source, R D Margulov, E K Selikhova and I Ya Furman, *Razvitie gazovoi promyshlennosti i analiz tekhniko-ekonomicheskikh pokazatelei (nauchno-ekonomicheskii obzor)*, Moscow, Ministerstvo gazovoi promyshlennosti, 1976.

During the period in which natural gas has made an increasing contribution to the Soviet energy balance, essentially the post-1960 period, there have been substantial changes not only in the geography of production but also of consumption. From 1961 to 1965 the growth in consumption in European Russia was accompanied by a corresponding rate of growth in production. However, during 1966-1970 the substantial part of European production was provided by operations in the North of the area, whereas the prime growth areas for consumption were the Ukraine, Volga, Transcaucasia, the Centre, the North-West and the South-West, as shown in the following table: (billion cubic metres).

Production Region	*1960*	*1965*	*1970*	*1975 Preliminary*
Total USSR	45.1	129.0	197.9	274.3
including RSFSR	23.9	75.6	117.3	160.5
North-West	1.8	6.6	9.5	13.4
Central	6.5	22.5	32.0	38.5
Central Chernozem	0.9	2.7	4.6	5.6
Volga	8.8	18.0	22.9	36.5
North Caucasus	4.4	11.2	15.3	18.2
Ural	0.5	10.2	26.2	38.7
Ukraine	14.6	34.6	49.6	58.3
Belorussia	–	2.3	2.9	4.0
Transcaucasia (Georgia, Azerbaidzhan, Armenia)	5.9	7.8	9.3	16.7

Source: *op cit*, p 8

Gas consumption increased by 40 per cent during the ninth Plan. The largest increases in individual sectors were recorded by the chemical industry (80 per cent), ferrous metallurgy (30 per cent), engineering and metalworking (30 per cent) and the non-ferrous metallurgical industry showed an increase of 150 per cent. In 1975, gas accounted for 79 per cent of fuel used in the production of ammonia (the base for fertilizer), 64 per cent of fuel in cement production and 35 per cent of fuel in copper production. A total of 126 billion cubic metres of gas was consumed in energy conversion (electricity and steam raising). The following table details the consumption of gas by sector of the economy, 1975 compared with 1970.

Gas Consumption Pattern 1970-1975
(billion cubic metres)

Consumption Sector	*1970*	*1975*	*1975 as % of 1970*
Total Soviet Consumption	190.8	264.5	139
Communal Use	25.3	34.0	134
Industrial Use	108.1	154.4	143
including Ferrous Metallurgy	28.0	35.4	126
Non-Ferrous Metallurgy	2.3	5.7	248
Engineering and Metalworking	19.1	24.2	127
Construction Materials	18.3	24.8	135
Oil and Gas	10.6	20.5	193
Fisheries	6.1	8.2	134
Light Industry	2.5	3.8	152
Other Industrial	8.3	9.0	108
Electricity Generation	52.7	67.8	129
Construction	1.0	1.7	170
Transport	0.7	1.3	186
Agriculture	0.7	1.9	271
Other	2.3	3.4	148

Source: *ibid*, p 14

From 1971 to 1975 gas consumption in the North-West, the Baltic states and Belorussia increased by 6.4 billion cubic metres, in the Central region by 8.9 billion, the Volga area by 13.6 billion, the Urals by 12.0 billion and in the Ukraine by 8.7 billion. However, concurrently with growth in consumption in European Russia, there has been a decline in production in the Ukraine and the North Caucasus which during the eighth Plan provided not only for demand in European Russia and the Central regions, but also in the North-West, the Baltic states and Belorussia. In the course of the ninth Plan this decline totalled 57 billion cubic metres, despite the installation of new productive capacity to offset depletion. The following table details the changes in gas consumption by region, 1975 compared with 1970.

Gas Consumption by Region 1970-1975
(billion cubic metres)

Economic Region	Consumption		Growth	
	1970	*1975*	*Absolute Increase*	*1975 as % increase on 1970*
North-West	9.5	13.4	3.9	41.1
Centre	32.0	38.5	6.5	20.3
Central Chernozem	4.6	5.6	1.0	21.7
Volga-Vyatskii	4.7	6.1	1.4	29.8
Volga	22.9	36.5	13.6	59.4
North Caucasus	15.3	18.2	2.9	19.0
Urals	26.2	38.7	12.5	47.7
Baltic states	2.8	4.2	1.4	50.0
Belorussia	2.9	4.0	1.1	37.9
Ukraine	49.6	58.3	8.7	17.5
Transcaucasia	9.3	16.7	7.4	79.6
Kazakhstan	3.3	7.6	4.3	130.3
Central Asia	13.4	23.9	10.5	78.4
Other Regions	1.4	8.7	7.3	521.0
Total	197.9	280.4	82.6	41.7

(NB figures include transportation loss)

Source: *ibid*, p 42

Appendix C

The Soviet System of Reserve Classification

Soviet reserve classifications for oil and gas do not readily correspond with those used in the West. Recent Western works have included details of the Soviet system:[1] for ease of reference the terms of the classification are detailed below. The information contained in this appendix is derived from a presentation by M Sh Modelevsky of the Laboratory of the Geology of Foreign Countries, Moscow, and V F Pominov of the Institute of World Economics and International Relations, Moscow, to the Conference on Energy Resources held in Laxenburg, Austria, in May 1975 under the auspices of the International Institute for Applied Systems Analysis (IIASA). The system of classification as given by Modelevsky and Pominov is used in most of the other socialist countries.

The classification divides known reserves into two main groups, those recoverable under present technical and economic conditions and those not recoverable under the same conditions. The group termed 'explored' ('razvedannye') reserves comprise classifications A, B and C1.

Category A reserves are calculated only during the time in which a given field is being exploited and when the factors influencing commercial production are known, such as total oil in place, chemical composition, seam pressure and the degree of gas saturation.

Category B comprises reserves in wells that have been extensively drilled and indicate estimated levels of production by tested flows from at least two wells. Knowledge of the chemical composition of the oil or gas and an estimate of the total exploitability of the field, adequate to justify the decision to develop, are required.

Category C1 comprises reserves in known fields, which contain a number of wells proved to be commercial. These reserves are normally found adjacent to deposits already classified in the A and B category, and where primary logging data indicates the likelihood of the reserves becoming commercially exploitable. It is noted that the level of accuracy experienced in estimating reserve levels is 10, 25 and 50 per cent for categories A, B and C1 respectively.

Category C2 relates to 'prospective' ('perspektivnye') reserves, found mainly in new exploration areas and calculated on the basis of logging data, and geological proximity to fields containing reserves in the higher categories.

It should be noted that reserves in new fields are usually reckoned in categories C1 and C2, and only occasionally in category B, immediately after drilling and testing the capacity of the first well to flow. Depending on the rate of success of development drilling, reclassification of reserves in category C1 is possible in a relatively short time.

Category D comprises 'predicted' ('prognoznye') reserves. In category D1 are found reserves in deposits at depths already reached in development drilling and the extent of which can be gauged from primary seismic data. No measure of the techno-commercial potential is assumed. Category D2 accounts for reserves that are believed to exist in regions with proven potential but where drilling has not been carried out, or where the depth of the likely deposit has not been reached in exploratory drilling carried out in existing hydrocarbon-bearing areas.[2]

It is stressed that the Soviet approach to reserve classification is primarily geological and that 'rather high' recovery factors, based on the ready availability of the most advanced technology, have been used. The economic factors have rarely been considered adequately. However, Modelevsky and Pominov did stress that there has recently been a shift in emphasis on the Soviet part to a 'more commercially oriented' system of classification. This includes setting limits to the degree of exploration regarded as sufficient to justify the decision to develop a given field, in order to avoid what may be excessive drilling costs incurred during the exploratory phase. On the question of comparability with Western classification systems, the authors stress the equal importance given in the West to economic considerations, and note that such an approach gives rise to underestimation, by Soviet standards, of ultimate explored reserves. The following tables, taken from the Modelevsky and Pominov paper, summarize the Soviet classification system (table A), and compare this system with others currently in use elsewhere in the hydrocarbon-producing world (table B).

References

1. (a) I F Elliot, *The Soviet Energy Balance*, London, Praeger, 1974, p 19. (b) J Bethkenhagen, *Bedeutung und Möglichkeiten des Ost-West-Handels mit Energierohstoffen*, (Deutches Institut für Wirtschaftsforschung, Sonderheft 104), Berlin, Duncker & Humblot, 1975, pp 29-30. (c) J Russell, *Energy as a Factor in Soviet Foreign Policy*, Farnborough, Saxon House, 1976, pp 221-222.

2. Note also that there is data on the content of reserve categories A, B and C in L M Umanskii, M M Umanskii, *Ekonomika neftyanoi i gazovoi promyshlennosti*, Moscow, Nedra, 1974, p 255.

Table A. General Soviet System of Reserve Classification

- Original potential resources
  - Remaining potential resources
    - Unexplored (undiscovered) reserves (C2 + D)
      - Forecast reserves (D)
        - In areas, horizons and depth ranges with established oil and gas bearing (D1)
        - In areas, deposits and depth ranges with expected oil and gas bearing (D2)
      - Prospective reserves (C2)
        - In some undrilled blocks of known fields
        - In prospective undrilled areas
  - Cumulative production
    - Current explored (discovered) reserves (A + B + C1)
      - In pools and blocks being developed (A + B)
      - In pools and blocks being explored (C1)

Table B. Comparison of Classification Systems Adopted in the USSR and in Other Countries

Groups of reserves in the US classification	Categories and groups from classification adopted in the USSR							
	USA and Canada	*India*	*Iran*	*Malaysia*	*France*	*Netherlands*	*West Germany*	*North African countries*
Proved	A, B partly C1	A, B	A, B	A, B	A, B partly C1	A, B partly C1	A, B partly C1	A, B, C1
Probable	—	C1, C2	C1	C1	C1 and partly C2 in known fields	mainly C1	C1 and sometimes C2 in known fields	—
Possible	D1, partly D2	—	C2 in known fields	C2 in known fields	C2 in prospective undrilled areas	C2 in known fields	C2 in prospective undrilled areas	—
Speculative	D2	D1	—	—	—	—	—	—

Selected Bibliography

A complete bibliography of the individual works cited in the text of this book would run to many pages. The purpose of this selected bibliography is to list the principal Western, mainly English-language, publications, the Comecon technical and economic journals and primary works on which the majority of the analysis is based. It is hoped that the detailed references in the text will be adequate to facilitate further research.

Principal Western Analyses of Comecon Energy Questions

Becker A S, 'Oil and the Persian Gulf in Soviet Foreign Policy', in Confino M, Shamir S (eds), *The USSR and the Middle East*, Jerusalem, Israel Universities Press, 1973, pp 173-214.

Bethkenhagen J, 'Die Zusammenarbeit der RGW-Länder auf dem Energiesektor', *Osteuropa Wirtschaft*, 2/1977, pp 63-80.

Bethkenhagen J, *Bedeutung und Möglichkeiten des Ost-West-Handels mit Energierohstoffen* (Deutsches Institut für Wirtschaftsforschung, Sonderheft 104), Berlin, Duncker & Humblot, 1975.

Bröll W, 'Die energetische Integration des RGW-Raumes', *Osteuropa Wirtschaft*, 3/1977, pp 26-49.

Campbell R W, *The Economics of Soviet Oil and Gas*, Baltimore, Johns Hopkins Press, 1968.

Campbell R W, *Trends in the Soviet Oil and Gas Industry*, Baltimore, Johns Hopkins Press, 1976.

Chesshire J, Huggett C, 'Primary Energy Production in the Soviet Union', *Energy Policy*, September 1975, pp 223-242.

Dienes L, 'Energy Prospects for Eastern Europe', *Energy Policy*, June 1976, pp 119-129.

Ebel R E, *Communist Trade in Oil and Gas*, New York, Praeger, 1970.

Elliot I F, *The Soviet Energy Balance*, New York, Praeger, 1974.

Gumpel W, *Die Energiepolitik der Sowjetunion*, Cologne, Wissenschaft und Politik, 1970.

Gumpel W, 'Sowjetunion: Erdöl und Nahostpolitik', *Aussenpolitik*, 11/1971, pp 670-681.

Jack E E, Lee J R, Lent H H, 'Outlook for Soviet Energy', in *The Soviet Economy in a New Perspective*, Washington DC, US Congress, Joint Economic Committee, 1976, pp 460-478.

Kaser M C, 'Oil and the Broader International Participation of IBEC', *International Currency Review*, 4/1974, pp 25-27.

Klinghoffer A J, *The Soviet Union and International Oil Politics*, New York, Columbia University Press, 1977.

Korda B, Moravcik I, 'The Energy Problem in Eastern Europe and the Soviet Union', *Canadian Slavonic Papers*, 3/1976, pp 1-14.

Lee J R, 'Petroleum Supply Problems in Eastern Europe', in *Reorientation and Commercial Relations of the Economies of Eastern Europe*, Washington DC, US Congress, Joint Economic Committee, 1974, pp 406-420.

NATO Economic Directorate, *Exploitation of Siberia's Natural Resources*, Brussels, NATO, 1974.

Park J D, 'OPEC and the Superpowers: An Interpretation', *Coexistence*, April 1976, pp 49-64.

Park J D, 'Oil and Gas in Comecon', in *The European Offshore Oil and Gas Yearbook 1976/1977*, London, Kogan Page, 1976, pp 257-263.

Park J D, 'Hydrocarbon Policy and Offshore Developments in Comecon', in *The Offshore Oil and Gas Yearbook 1978/79, UK and Continental Europe*, London, Kogan Page, 1978, pp 193-200.

Polach J G, 'The Development of Energy in East Europe', in *Economic Developments in Countries of Eastern Europe*, Washington DC, US Congress, Joint Economic Committee, 1970, pp 348-433.

Russell J L, 'Energy Considerations in Comecon Policies', *The World Today*, February 1976, pp 39-48.

Russell J L, *Energy as a Factor in Soviet Foreign Policy*, Farnborough, Saxon House, 1976.

Shimkin D B, *The Soviet Mineral-Fuels Industries 1928-1958: A Statistical Survey*, Washington DC, US Government Printing Office, 1962.

Slocum M, 'Soviet Energy – An Internal Assessment', *Technology Review*, October/November 1974, pp 17-33.

Principal Economic and Technical Journals

Ekonomika neftyanoi promyshlennosti (Moscow)
Energetica (Bucharest)
Energieanwendung (Leipzig)
Energietechnik (Leipzig)
Gazovaya promyshlennost' (Moscow)
Khimiya i tekhnologiya topliv i masel (Moscow)
Mine, petrole şi gaze (Bucharest)
Nafta (Krakow)
Neftepererabotka i neftekhimiya (Moscow)
Neftepromyslovoe delo (Moscow)
Neftyanik (Moscow)
Neftyanoe khozyaistvo (Moscow)
Promyshlennaya energetika (Moscow)
Stroitel'stvo truboprovodov (Moscow)
Wiadomosci naftowe (Krakow)

Primary Soviet Sources

Bakirov A A, Ryabukhin G E, *Neftegazonosnye provintsii i oblasti SSSR*, Moscow, Nedra, 1969.

Brenner M M, *Ekonomika neftyanoi i gazovoi promyshlennosti*, Moscow, Nedra, 1968.

Feigin M B, *Neftyanye resursy – metodika ikh issledovaniya i otsenki*, Moscow, Nedra, 1974.

Karyagin I D, *Ekonomicheskie problemy razvitiya neftyanoi promyshlennosti Zapadnoi Sibiri*, Moscow, Nedra, 1975.

Kortunov A K, *Gazovaya promyshlennost' SSSR*, Moscow, Nedra, 1967.

Kozlov I D, Shmakova E K, *Sotrudnichestvo stran-chlenov SEV v energetike*, Moscow, Nauka, 1973.

Kozyrev V M, *Renta, tsena i khozrashchet v neftyanoi promyshlennosti*, Moscow, Nedra, 1972.

Ladygin B N, Motorin I F, *Problemy sotrudnichestva stran SEV v razvitii toplivnoi-syr'evoi bazy*, Moscow, Nedra, 1968.

L'vov M A, *Resursy prirodnogo gaza SSSR*, Moscow, Nedra, 1969.

Luzin V I, *Ekonomicheskaya effektivnost' i planirovanie kapital'nykh vlozhenii i osnovnykh fondov v neftyanoi promyshlennosti*, Moscow, Nedra, 1974.

Makarov A A, Melent'ev L A, *Metody issledovaniya i optimizatsii energeticheskogo khozyaistva*, Novosibirsk, Nauka, 1973.

Maksakovskii V P, *Toplivnye resursy sotsialisticheskikh stran Evropy*, Moscow, Nedra, 1968.

Maksakovskii V P, *Toplivnaya promyshlennost' sotsialisticheskikh stran Evropy*, Moscow, Nedra, 1975.

Margulov R D, Selikhova E K, Furman I Ya, *Razvitie gazovoi promyshlennosti i analiz tekhniko-ekonomicheskikh pokazatelei*, Moscow, Ministerstvo gazovoi promyshlennosti, 1976.

Mel'nikov N V, *Mineral'noe toplivo*, Moscow, Nedra, 1971.

Mel'nikov N V, *Toplivno-energeticheskie resursy SSSR*, Moscow, Nauka, 1971.

Pavlenko A S, Nekrasova A M (eds), *Energetika SSSR v 1971-1975 gody*, Moscow, Energiya, 1972.

Probst A E, *Razvitie toplivnoi bazy raionov SSSR*, Moscow, Nedra, 1968.

Probst A E, Mazover Ya A (eds), *Razvitie i razmeshchenie toplivnoi promyshlennosti*, Moscow, Nedra, 1975.

Shelest V A, *Regional'nye energo-ekonomicheskie problemy SSSR*, Moscow, Nedra, 1975.

Sokolov G D, *Kapital'noe stroitel'stvo v neftyanoi promyshlennosti*, Moscow, Nedra, 1973.

Tomashpol'skii L M, *Neft' i gaz: problemy i prognozy*, Moscow, Nedra, 1975.

Umanskii L M, Umanskii M M, *Ekonomika neftyanoi i gazovoi promyshlennosti*, Moscow, Nedra, 1975.

Urinson G S, et al, *Ekonomika razrabotki gazovykh mestorozhdenii*, Moscow, Nedra, 1973.

Ushakov S S, *Tekhniko-ekonomicheskie problemy transporta topliva*, Moscow, Transport, 1972.

Index

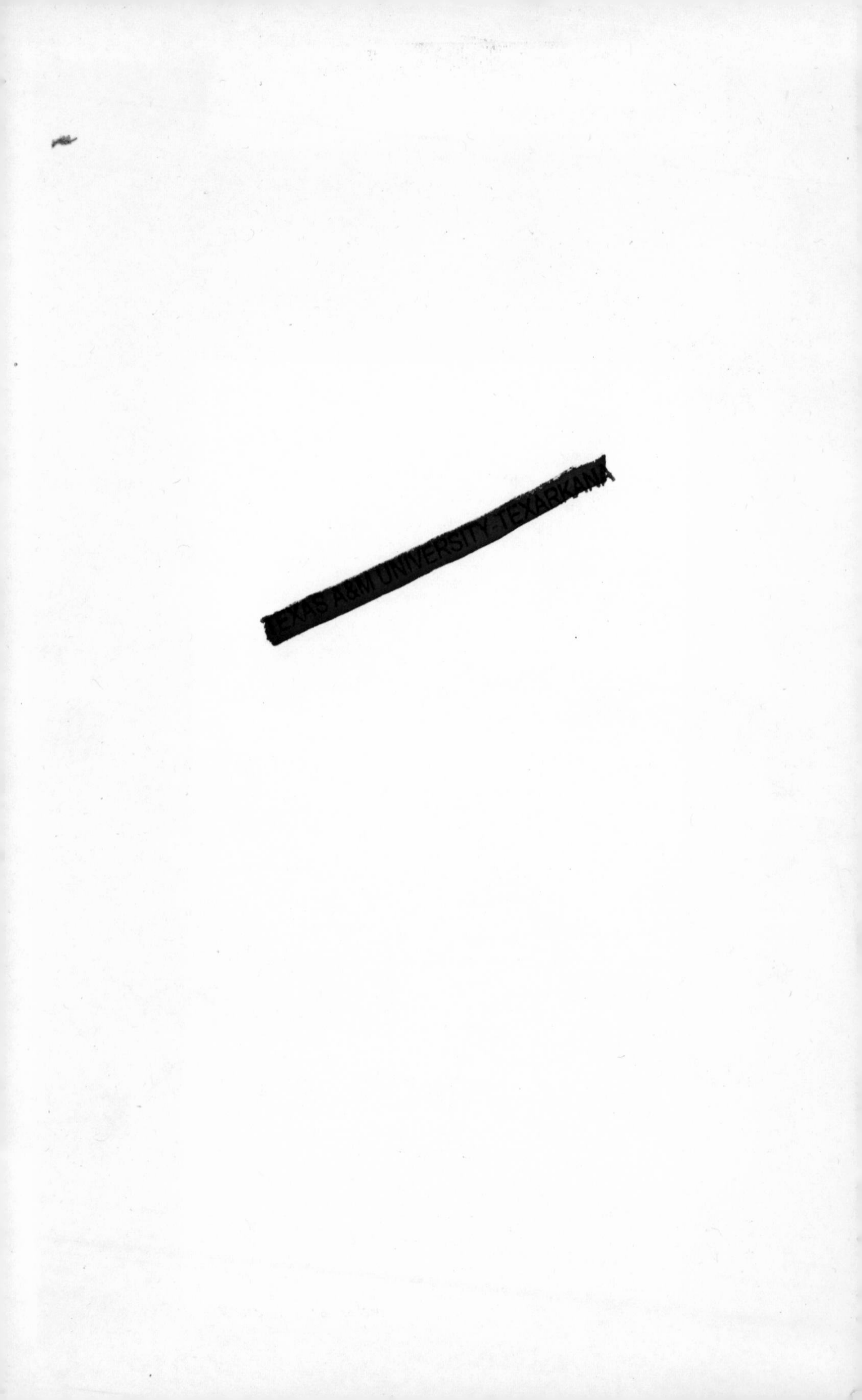